Math Connects

LENNA

Homework and Problem-Solving Practice Workbook

Course 2

Also available online at

connectED.mcgraw-hill.com

Mc Graw Hill **Glencoe**

To the Teacher These worksheets are the same as those found in the Chapter Resource Masters for Glencoe's *Math Connects*, Course 2. The answers to these worksheets are available at the end of each Chapter Resource Masters booklet.

The McGraw·Hill Companies

 Glencoe

Copyright © by The McGraw-Hill Companies, Inc. All rights reserved. Except as permitted under the United States Copyright Act, no part of this publication may be reproduced or distributed in any form or by any means, or stored in a database or retrieval system, without prior permission of the publisher.

Send all inquiries to:
Glencoe/McGraw-Hill
8787 Orion Place
Columbus, OH 43240-4027

ISBN: 978-0-07-895137-4
MHID: 0-07-895137-2

Homework and Problem-Solving Practice Workbook, Course 2

Printed in the United States of America.

13 14 PRS 19 18 17 16 15 14 13

CONTENTS

Chapter 7 Linear Functions

Multi-Part Lesson 1 **Rates and Functions**

Multi-Part Lesson 2 **Slope**

Multi-Part Lesson 3 **Variation**

Chapter 8 Probability and Predictions

Multi-Part Lesson 1 **Probability**

Multi-Part Lesson 2 **Compound Events**

Multi-Part Lesson 3 **Predictions**

Chapter 9 Statistical Displays

Multi-Part Lesson 1 **Measures of Central Tendency**

Multi-Part Lesson 2 **Measures of Variation**

Multi-Part Lesson 3 **Statistical Displays**

Multi-Part Lesson 4 **More Statistical Displays**

Chapter 10 Volume and Surface Area

Multi-Part Lesson 1 **Volume**

Multi-Part Lesson 1

PART A

Homework Practice

Powers and Exponents

Write each power as a product of the same factor.

1. 5^7

2. 2^4

3. 7^2

4. 10^5

5. 3^3

6. 6^8

7. *four to the eighth power*

8. *eight cubed*

9. *ten squared*

Write each product in exponential form.

10. $9 \cdot 9 \cdot 9 \cdot 9 \cdot 9 \cdot 9$

11. $1 \cdot 1 \cdot 1 \cdot 1 \cdot 1$

12. $2 \cdot 2 \cdot 2 \cdot 2 \cdot 2 \cdot 2 \cdot 2$

13. $6 \cdot 6 \cdot 6 \cdot 6 \cdot 6 \cdot 6 \cdot 6 \cdot 6 \cdot 6$

14. $5 \cdot 5$

15. $4 \cdot 4 \cdot 3 \cdot 3 \cdot 3 \cdot 3$

Evaluate each expression.

16. 4^3

17. 1^{11}

18. 2^5

19. 10^3

20. 9^3

21. 8^1

22. *five to the fourth power*

23. $\frac{2}{3}$ *squared*

24. *zero to the sixth power*

25. Write $3 \cdot 3 \cdot 3 \cdot 4 \cdot 4 \cdot 4 \cdot 4$ in exponential form.

Order the following powers from least to greatest.

26. $7^2, 5^3, 3^4, 2^5$

27. $4^3, 1^{13}, 12^2, 8^3$

28. $3^9, 5^7, 7^5, 9^3$

29. **INTERACTIVE MAPS** Mansi is using an interactive map on her computer that allows her to zoom in or zoom out. Each time she zooms out the scale of the map increases by a power of ten. If she zooms out four times the scale is 10^4 times greater. Write this number in standard form.

30. **BACTERIA** A lab technician observed 5 bacteria growing in a lab dish. One hour later he observed 25 bacteria. Every hour he notices about 5 times as many as the hour before. After several hours of observation, he determined the lab dish had 5^9 bacteria. Use a calculator to find the number in standard form that represents the bacteria in the lab dish.

Get ConnectED *For more practice, go to www.connected.mcgraw-hill.com.*

Multi-Part Lesson 1

PART A

Problem-Solving Practice

Powers and Exponents

---

1. BUDGETS Shiangtai's office has a budget of about $10 \cdot 10 \cdot 10 \cdot 10 \cdot 10 \cdot 10 \cdot 10$ dollars. Write this amount in exponential form.

2. ANIMALS The African bush elephant is the largest land animal and weighs about $2 \cdot 2 \cdot 2$ tons. Write this amount in exponential form.

---

3. VOLUME To find the volume of a rectangular box, you multiply the length times the width times the height. In a cube, all sides are the same length. If the cube has length, width, and height of 6 inches, write the volume as a product. Then write it in exponential form.

4. SCIENCE A certain type of cell doubles every hour. If you start with one cell, at the end of one hour you would have 2 cells, at the end of two hours you have 4 cells, and so on. The expression $2 \times 2 \times 2 \times 2 \times 2$ tells you how many cells you would have after five hours. Write this expression in exponential form; then evaluate it.

---

5. MATH Evaluate 5^4 and 25^2. What do you notice?

6. PREFIXES Many prefixes are used in mathematics and science. The prefix *giga* in gigameter represents 1,000,000,000 meters. Write this number as a power of ten.

---

7. LIBRARY The school library contains 9^4 books. How many books are in the school library?

8. HOT DOGS The concession stand at the county fair sold 6^3 hot dogs on the first day. How many hot dogs did they sell?

Course 2 • Expressions and Patterns

Homework Practice

Numerical Expressions

Evaluate each expression. Justify each step.

1. $(2 + 9) \times 4$

2. $8 - (5 + 2)$

3. $(15 \div 3) + 7$

4. $(14 + 7) \div 7$

5. $5 \cdot 6 - 12 \div 4$

6. $8 \div 2 + 8 - 2$

7. $16 - 8 \div 2 + 5$

8. $15 - 3 \cdot 5 + 7$

9. 7×10^3

10. $2 \times 5^2 + 6$

11. $7 \cdot 2^3 - 9$

12. $27 \div 3 \times 2 + 4^2$

13. $6^3 - 12 \times 4 \cdot 3$

14. $(15 - 3) \div (8 + 4)$

15. $(9 - 4) \cdot (7 - 7)$

16. $8 + 3(5 + 2) - 7 \cdot 2$

17. $5(6 - 1) - 4 \cdot 6 \div 3$

18. $(5 + 7)^2 \div 12$

19. $12 \div (8 - 6)^2$

20. $(7 + 2)^2 \div 3^2$

21. $(11 - 9)^2 \cdot (8 - 5)^2$

22. $64 \div 8 - 3(4 - 3) + 2$

23. $8 \times 5.1 - (4.1 + 1.4) + 7.1$

For Exercises 24 and 25, write an expression for each situation. Then evaluate the expression to find the solution.

24. LAWN AREA The Solomons need to find the area of their front and side yards since they want to reseed the lawn. Both side yards measure 3 meters by 10 meters, while the front yard is a square with a side of 9 meters. They do not need to reseed a portion of the front yard covering 16 square meters where a flower bed is located. What is the area of the yard that the Solomons want to reseed?

25. COMMUNITY SERVICE Jariah volunteers at the hospital during the week. She volunteers 3 hours on Monday and Thursday, 4 hours on Saturday and Sunday, and 2 hours on Tuesday. How many hours does Jariah volunteer at the hospital during the week?

Get ConnectED *For more practice, go to* www.connected.mcgraw-hill.com.

Problem-Solving Practice
Numerical Expressions

---

1. FOOTBALL The middle school team scored three field goals worth three points each and two touchdowns with extra points worth seven points each. Write a numerical expression to find the team's score. Then evaluate the expression.

2. BOOKS Juan goes to the school book fair where paperback books are $1.50 and hardback books are $3.00. Juan buys 5 paperback and 2 hardback books. Write a numerical expression to find how much Juan paid for the books. Then evaluate the expression.

---

3. GEOMETRY The perimeter of a hexagon is found by adding the lengths of all six sides of the hexagon. For the hexagon below write a numerical expression to find the perimeter. Then evaluate the expression.

4. MONEY Aisha bought school supplies consisting of 6 spiral notebooks costing $0.39 each, 2 packages of pencils at $0.79 each, and a 3-ring binder for $1.99. Write an expression to find the total amount Aisha spent on school supplies. Then evaluate the expression.

---

5. REASONING Use the order of operations and the digits 2, 4, 6, and 8 to create an expression with a value of 2.

6. NUMBER SENSE Without parentheses, the expression $8 + 30 \div 2 + 4$ equals 27. Rewrite the expression with parentheses so that it equals 13; then 23.

---

7. MONEY Tyrone bought 5 postcards at $0.55 each and a set of postcards for $1.20. Write an expression to find the total amount Tyrone spent on postcards. Then evaluate the expression.

8. DINING Mr. Firewalks took his family out to eat. They ordered 3 meals costing $8.99 each, 2 sodas at $1.50 each, and 1 glass of tea for $1.25. Write an expression to find the total amount the Firewalks family spent on dinner before taxes and tip. Then evaluate the expression.

Homework Practice

Algebraic Expressions

Evaluate each expression if $r = 5$, $s = 2$, $t = 7$, and $u = 1$.

1. $s + 7$ **2.** $9 - u$ **3.** $3t + 1$

4. $5r - 4$ **5.** $t - s$ **6.** $u + r$

7. $11t - 7$ **8.** $6 + 3u$ **9.** $4r - 10s$

10. $3u^2$ **11.** $2t^2 - 18$ **12.** $r^2 + 8$

13. $\dfrac{s}{2}$ **14.** $\dfrac{30}{r}$ **15.** $\dfrac{(3 + u)^2}{8}$

Evaluate each expression if $a = 4.1$, $b = 5.7$, and $c = 0.3$.

16. $a + b - c$ **17.** $10 - (a + b)$ **18.** $b - c + 2$

19. MOON The expression $\dfrac{w}{6}$ gives the weight of an object on the Moon in pounds with a weight of w pounds on Earth. What is the weight of a space suit on the Moon if the space suit weighs 178.2 pounds on Earth?

20. Complete the table.

Pounds (p)	Ounces (16p)
1	16
2	32
3	
4	
5	

Get ConnectED *For more practice, go to* www.connected.mcgraw-hill.com.

Problem-Solving Practice

Algebraic Expressions

---

1. FIELD TRIP The seventh grade math classes are going on a field trip. The field trip will cost $7 per student. Write an expression to find the cost of the field trip for s students. What is the total cost if 26 students go on the trip?

2. SOCCER Jason earns $20 per game as a referee in youth soccer games. Write an expression to find how much money Jason will earn for refereeing any number of games. Let n represent the number of games Jason has refereed. How much will he earn for refereeing 6 games?

---

3. PROFIT The expressions $c - e$, where c stands for the money collected and e stands for the expenses, is used to find the profit from a basketball concession. If $500 was collected and expenses were $150, find the profit for the concession.

4. SAVINGS Kata has a savings account that contains $230. She decides to deposit $5 each month from her monthly earnings for baby-sitting after school. Write an expression to find how much money Kata will have in her savings account after x months. Let x represent the number of months. Then find out how much she will have in her account after 1 year.

---

5. MONEY Mr. Wilson has $2,500 in his savings account and m dollars in his checking account. Write an expression that describes the total amount that he has in both accounts.

6. ANIMALS Write an expression to represent the total number of legs on h horses and c chickens. How many legs are there in 5 horses and 6 chickens?

---

7. T-SHIRTS The band wants to order T-shirts. The T-shirts cost $15 each plus a shipping fee of $10. Write an expression to find the total cost of c T-shirts.

8. TEMPERATURE The expression $\frac{9}{5}C + 32$, where C stands for temperature in degrees Celsius, is used to convert Celsius to Fahrenheit. If the temperature is 20 degrees Celsius, find the temperature in degrees Fahrenheit.

NAME _____ DATE _____ PERIOD _____

Homework Practice

Properties

Use the Distributive Property to re-write each expression. Then evaluate it.

1. $4(5 + 7)$ **2.** $6(3 + 1)$ **3.** $(10 + 8)2$

4. $5(8 - 3)$ **5.** $7(4 - 1)$ **6.** $(9 - 2)3$

Evaluate each expression mentally. Justify each step.

7. $7 + (6 + 3)$ **8.** $23 \cdot 15$ **9.** $0 + 1$

10. $3 \cdot 4 + 3 \cdot 7$ **11.** 8×1 **12.** $11 + 0$

13. $5(9 + 1)$ **14.** $4 \cdot (7 \cdot 1)$ **15.** $(6)7 + (2)7$

Use one or more properties to rewrite each expression as an equivalent expression that does not use parentheses.

16. $(b + 3) + 6$ **17.** $7(5x)$ **18.** $4(a + 4)$

19. $7 + (3 + t)$ **20.** $(2z)0$ **21.** $(9 + k)5$

22. $8(y - 5) + y$ **23.** $(h + 2)3 - 2h$

24. GROCERY A grocery store sells an imported specialty cheesecake for $11 and its own store-baked cheesecake for $5. Use the Distributive Property to mentally find the total cost for 6 of each type of cheesecake.

25. CHECKING ACCOUNT Mr. Kenrick balances his checking account statement each month two different ways as shown by the equation, $(b + d) - c = b + (d - c)$, where b is the previous balance, d is the amount of deposits made, and c is the amount of checks written. Name the property that Mr. Kenrick uses to double check his arithmetic.

26. SPEED A train is traveling at a speed of 65 miles per hour. The train travels for one hour. What property is used to solve this problem as shown by the statement $65 \cdot 1 = 65$?

Get ConnectED *For more practice, go to* www.connected.mcgraw-hill.com.

Problem-Solving Practice

Properties

1. MUSIC Mr. Escalante and Mrs. Turner plan to take their music classes to a musical revue. Tickets cost $6 each. Mr. Escalante's class needs 22 tickets, and Mrs. Turner's class needs 26 tickets. Use the Distributive Property to write a sentence to express how to find the total cost of tickets in two ways.

2. SAVINGS Mrs. Perez was looking at her bank account statement. She noticed that her beginning balance was $500, and she had added nothing to her account. What was the ending balance on her statement? What property did you apply?

3. ADDITION Mr. Brooks was working on addition using dominos with a group of 1st graders. When picking the domino with 3 dots on one end and 5 dots on the other, some students read. "3 plus 5 equal 8" while others read it as "5 plus 3 equals 8." What property were these children using? Explain.

4. AREA Aleta noticed that for the rectangle below she could either multiply 2 times 3 or 3 times 2 to get its area of 6 square inches. What property allows her to do this?

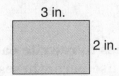

3 in.

2 in.

5. NUMBER CUBES Students in Mr. Rivas' class were practicing their multiplication skills by rolling three 6-sided number cubes. Wapi rolled a 2, a 3, and a 5 on his roll. He multiplied the three numbers as follows using the order of operations: $(2 \times 3) + 5 = 30$. Write another way Wapi could have performed the multiplication without changing the order of the numbers. State the property you used.

6. FACTS Bik was working on memorizing her multiplication facts. She noticed that anytime she multiplied a number by 1, she got the same number she started with. What property allows this to be true?

7. MONEY Mei was trying to figure out the cost of 4 boxes of cereal for $2.25 each. Write a sentence to show Mei an easy way to do her calculations. What property did you apply to help her?

8. WALKING Jacob walked 3 blocks to Ping's house, then 5 blocks to Jamal's house. Write a sentence to show that the distance from Ping's to Jamal's is the same as the return walk home. Name the property illustrated in your sentence.

Multi-Part Lesson **2**
PART **A**

Homework Practice

Problem-Solving Investigation: Look for a Pattern

Use the *look for a pattern* strategy to solve Exercises 1–4.

1. **READING** Shayna is reading a new novel. The last three nights she has read 25, 31, and 37 pages. If she continues reading in this pattern, how many pages of the book can she expect to have read after the sixth night?

2. **TEMPERATURE** The table shows the daily high temperature for a city for the past four days. If the patttern continues, what would you expect the high temperature to be for the next two days?

Day	Temperature (°F)
Sun.	72
Mon.	73
Tues.	75
Wed.	78

3. **NUMBERS** What are the next three numbers in the pattern below?
138, 113, 88, ____, ____, ____

4. **TYPING** Parker is taking a typing class. His scores on his timed typing tests are 18, 20, and 24 words per minute. Parker has two more timed tests to take in the course. If the pattern continues, how many words per minute can Parker expect to be able to type at the end of the course?

Use any strategy to solve Exercises 5–8.

5. **DANCE** The cheerleaders are practicing a dance routine in which all 36 of them need to be in a triangular formation. There will be two more cheerleaders in each row than the previous row. How many rows will be in the formation?

6. **GEOMETRY** Draw the next two figures in the pattern shown below.

7. **PRECIPITATION** The table shows the average monthly precipitation for Seattle, Washington. About how much precipitation can Seattle expect to receive during March through August? For the whole year?

Average Monthly Precipitation for Seattle, Washington (in.)					
Jan.	5.1	May.	1.7	Sept.	1.6
Feb.	3.7	June.	1.4	Oct.	3.0
Mar.	3.3	July.	0.7	Nov.	5.1
Apr.	2.2	Aug.	0.9	Dec.	5.4

8. **CAKE** Tiffany is cutting a rectangular cake for a party. She needs 30 equal-sized pieces to serve all the guests. How many cuts will Tiffany need to make in the cake?

Get ConnectED *For more practice, go to* www.connected.mcgraw-hill.com.

Problem-Solving Practice

Problem-Solving Investigation: Look for a Pattern

1. WAGES The table shows the amount of annual pay raise Miss Jones received the last three years. If the pattern in her pay continues, how much can she expect her pay increase to be five years from now?

Year	Annual Raise ($)
Two Years Ago	500
Last Year	1,000
Current year	1,500

2. BABY SITTING For the last five weeks, Sahara has baby sat 4, 5, 7, 8, and 10 hours each week. If the pattern continues, how many total hours will she have baby sat in 10 weeks?

3. GEOMETRY Draw the next two figures in the pattern shown below.

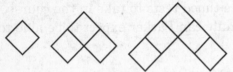

4. EXERCISE A trainer is recording a client's progress each week. The table shows the client's weight each week for the first four weeks of the program. If the pattern continues, how much total weight can he expect to lose after following the program for 12 weeks?

Week	1	2	3	4
Weight (lb)	145	142	139	136

5. CLUBS Attendance at the last three foreign language club meetings has been 24, 20, and 16 students. If attendance continues to change in this pattern, how many students can be expected to attend the next meeting?

6. STUDYING For his history exam tomorrow, Zachary has studied for 2 hours and 40 minutes. This is 10 minutes more than twice the amount of time he spent studying for his last exam. How many more minutes did Zachary study for his history exam than his last exam?

Multi-Part Lesson 2

PART B

Homework Practice

Sequences

Describe the relationship between the terms in each arithmetic sequence. Then write the next three terms in each sequence.

1. 0, 5, 10, 15, …

2. 1, 3, 5, 7, …

3. 18, 27, 36, 45, …

4. 7, 19, 31, 43, …

5. 8, 18, 28, 38, …

6. 25, 26, 27, 28, …

7. 0.4, 0.8, 1.2, 1.6, …

8. 3.7, 3.7, 3.7, 3.7, …

9. 5.1, 6.2, 7.3, 8.4, …

10. 17, 31, 45, 59, …

11. 30, 50, 70, 90, …

12. 14, 41, 68, 95, …

In a *geometric sequence*, each term is found by multiplying the previous term by the same number. Write the next three terms of each geometric sequence.

13. 5, 10, 20, 40, …

14. 3, 9, 27, 81, …

15. 2, 8, 32, 128, …

NUMBER SENSE Find the 40th term in each arithmetic sequence.

16. 4, 8, 12, 16, …

17. 13, 26, 39, 52, …

18. 6, 12, 18, 24, …

19. GEOMETRY The lengths of the sides of a 6-sided polygon are an arithmetic sequence. The length of the shortest side is 3 meters. If the length of the next longer side is 5 meters, what is the length of the longest side?

20. FREE FALLING OBJECT A free falling object increases speed by a little over 22 miles per hour each second. The arithmetic sequence 22, 44, 66, …, represents the speed after each second, in miles per hour, of a dropped object. How fast is a rock falling after 8 seconds if it is dropped over the side of a cliff?

Get ConnectED *For more practice, go to* www.connected.mcgraw-hill.com.

Problem-Solving Practice

Sequences

---

1. NUMBERS The multiples of two form a sequence as follows: 2, 4, 6, 8, 10, 12, 14, 16, …. Describe the sequence you see? What about the multiples of three? Four? Five?

2. OLYMPICS The summer Olympics occur every four years. If the last summer Olympics happened in 2008, when are the next three times that it will occur? Describe the sequence the Olympic years form.

---

3. BABY-SITTING Tonya charges $3.50 per hour to baby-sit. The sequence $3.50, $7.00, $10.50, $14.00, … represents how much she charges for each subsequent hour. For example, $10.50 is the third term that represents how much she charges for 3 hours. What are the next three terms in the sequence? How much does she charge for 7 hours of baby-sitting?

4. RECTANGLES Suppose you start with 1 rectangle and then divide it in half. You now have 2 rectangles. You divide each of these in half, and you have 4 rectangles. The sequence for this division is 1, 2, 4, 8, 16, . . . rectangles after each successive division. Describe the sequence that results?

---

5. BACTERIA Three bacteria are in a dish. Each hour the number of bacteria multiplies by four. If at the end of the first hour there are 12 bacteria, how many bacteria are there at the end of the next three hours? Describe the sequence that results?

6. ENROLLMENT The enrollment at Grove Middle School is expected to increase by 40 students each year for the next 5 years. If their current enrollment is 600 students, find their enrollment after each of the next 5 years.

---

7. SALARY Mrs. Malone's current salary is $15,000. She expects it to increase $1,000 per year. Write the first 6 terms of a sequence that represents her salary. The first term should be her current salary. What does the sixth term represent?

8. FIBONACCI The Fibonacci sequence is named after Leonardo Fibonacci who first explored it. Look at the Fibonacci sequence below and describe its pattern. 1, 1, 2, 3, 5, 8, 13, 21, 34, …

Multi-Part Lesson 3

PART B

Homework Practice

Squares and Square Roots

Find the square of each number.

1. 2

2. 8

3. 10

4. 11

5. 15

6. 25

Find each square root.

7. $\sqrt{64}$

8. $\sqrt{121}$

9. $\sqrt{169}$

10. $\sqrt{0}$

11. $\sqrt{81}$

12. $\sqrt{289}$

13. $\sqrt{900}$

14. $\sqrt{1}$

15. $\sqrt{484}$

16. **PACKAGING** An electronics company uses three different sizes of square labels to ship products to customers. The area of each type of label is shown in the table.

Labels	
Type	**Area**
Priority	100 cm²
Caution	225 cm²
Address	144 cm²

   a. What is the length of a side for each label?

   b. How much larger is the Caution label than the Address label?

17. **RECREATION** A square hot tub is outlined by a 2-foot wide tile border. In an overhead view, the area of the hot tub and the border together is 144 square feet. What is the length of one side of the hot tub itself?

Get ConnectED *For more practice, go to* www.connected.mcgraw-hill.com.

Problem-Solving Practice

Squares and Square Roots

1. FERTILIZER Bryce bought a bag of lawn fertilizer that will cover 400 square feet. What are the dimensions of the largest square plot of lawn that the bag of fertilizer will cover?

2. GEOMETRY The area A of a circle in square feet with a radius r in feet is given approximately by the formula $A = 3.14r^2$. What is the approximate area of a circle with a radius of 3 feet?

3. MOTION The time t in seconds for an object dropped from a height of h feet to hit the ground is given by the formula $t = \sqrt{\dfrac{2h}{32}}$. How long will it take an object dropped from a height of 500 feet to hit the ground? Round to the nearest tenth.

4. PACKAGING A cardboard envelope for a compact disc is a square with an area of 171.61 square centimeters. What are the dimensions of the envelope?

5. GEOGRAPHY Refer to the squares below. They represent the approximate areas of California, Alabama, and Nebraska. Find the area of Alabama.

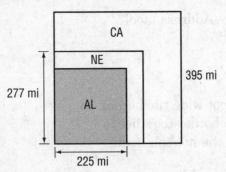

6. GEOGRAPHY Use the figure in Exercise 5. How much larger is California than Nebraska?

Homework Practice

Estimate Square Roots

Estimate each square root to the nearest whole number.

1. $\sqrt{8}$ 2. $\sqrt{19}$ 3. $\sqrt{47}$ 4. $\sqrt{70}$

5. $\sqrt{91}$ 6. $\sqrt{125}$ 7. $\sqrt{150}$ 8. $\sqrt{389}$

9. $\sqrt{2,468}$ 10. $\sqrt{899}$ 11. $\sqrt{4,840}$ 12. $\sqrt{8,080}$

13. $\sqrt{6}$ 14. $\sqrt{21}$ 15. $\sqrt{53}$ 16. $\sqrt{79}$

17. $\sqrt{190}$ 18. $\sqrt{624}$ 19. $\sqrt{427}$ 20. $\sqrt{3,178}$

21. **ALGEBRA** Estimate $\sqrt{a+b}$ to the nearest whole number if $a = 24$ and $b = 38$.

22. **ALGEBRA** Estimate $\sqrt{x-y}$ to the nearest whole number if $x = 10$ and $y = 4.5$

23. **QUILTING** A queen-size quilt in the shape of a square has an area of 51 square feet. What is the approximate length of one side of the quilt?

24. **PENDULUM** The formula below can be used to estimate the time it takes for a pendulum to swing back and forth once. Use the formula to find the approximate time it takes for a pendulum with a length of 0.8 meter to swing back and forth once.

$$T = 2 \times \sqrt{L}$$
- $T = $ time (seconds)
- $L = $ length (meters)

25. **ARCHITECTURE** A mansion in Ponte Vedra has a square front door with an area of 95 square feet. What is the approximate length of one side of the door?

Get ConnectED *For more practice, go to* www.connected.mcgraw-hill.com.

Problem-Solving Practice

Estimate Square Roots

---

1. **GEOMETRY** The diameter d of a circle with area A is given by the formula $d = \sqrt{\dfrac{4A}{\pi}}$ What is the diameter of a circle with an area of 56 square inches? Use 3.14 for π and round to the nearest whole number.

2. **FENCING** Hope wants to buy fencing to enclose a square garden with an area of 500 square feet. How much fencing does Hope need to buy? Round to the nearest whole number.

---

3. **OCEANS** The speed v in feet per second of an ocean wave in shallow water of depth d in feet is given by the formula $v = \sqrt{32d}$. What is the speed of an ocean wave at a depth of 10 feet? Round to the nearest whole number.

4. **LIGHTING** A new flashlight has a beam whose width w at a distance d from the flashlight is given by the formula $w = 1.2\sqrt{d}$. What is the width of the beam at a distance of 30 feet? Round to the nearest whole number.

---

5. **SOUND** The speed of sound in air c in meters per second at a temperature T in degrees Celsius is given approximately by the formula $c = \sqrt{402(T+273)}$. What is the speed of sound in air at a temperature of 25 degrees Celsius? Round to the nearest whole number.

6. **PROJECTILES** The muzzle velocity v in feet per second necessary for a cannon to hit a target x feet away is estimated by the formula $v = \sqrt{32x}$. What muzzle velocity is required to hit a target 3,000 feet away? Round to the nearest whole number.

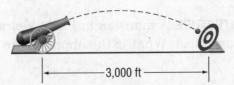

3,000 ft

Multi-Part Lesson 1
PART B

Homework Practice

Integers and Absolute Value

Write an integer for each situation.

1. a profit of $12

2. 1,440 feet below sea level

3. 22°F below zero

4. a gain of 31 yards

Graph each set of integers on a number line.

5. {−5, 0, 5}

6. {−3, −2, 1, −4}

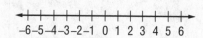

Evaluate each expression.

7. $|-11|$

8. $|-5| + 8$

9. $|-4| - |-4|$

10. $|12| \div 2 \times |-5|$

11. $|-4| + 7 - |3|$

12. $9 + |-6| \div |-3|$

13. HEALTH A veterinarian recommends that a St. Bernard lose weight. Write an integer to describe the dog losing 25 pounds.

14. GEOGRAPHY Mount Kilimanjaro is the highest peak in Africa. Write an integer to represent the elevation of Mount Kilimanjaro at 5,895 meters above sea level.

15. ECONOMY Gasoline prices occasionally fluctuate during a two-month period of time. Prices increased 34 cents per gallon during the month of April and decreased 17 cents per gallon during the month of May. Which situation has the greater absolute value? Explain.

Get ConnectED *For more practice, go to* www.connected.mcgraw-hill.com.

Multi-Part Lesson 1

PART B

Problem-Solving Practice

Integers and Absolute Value

1. DEATH VALLEY The lowest point in the United States is Death Valley in California. Its altitude is 282 feet below sea level. Write an integer to represent the altitude of Death Valley.	**2. RAIN** A meteorologist reported that in the month of April there were 3 inches more rainfall than normal. Write an integer to represent the amount of rainfall above normal in April.
3. ARCHIMEDES A famous mathematician and physicist named Archimedes was born in 287 B.C. Write an integer to express the year of his birth.	**4. TEMPERATURE** In our world's tropical rain forests, the average temperature of every month is 64 degrees above zero or higher. Write an integer to express this temperature.
5. STOCK MARKET A certain stock gained 5 points in one day and lost 4 points the next day. Which situation has the greater absolute value? Explain.	**6. ALTITUDE** An airplane pilot changed his altitude by 100 meters. Describe what this could mean.

Multi-Part **1**
Lesson

PART **C**

Homework Practice

The Coordinate Plane

Write the ordered pair that corresponds to each point graphed at the right. Then state the quadrant or axis on which each point is located.

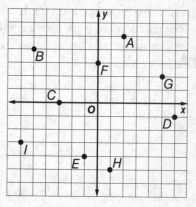

1. A   **2.** B   **3.** C

4. D   **5.** E   **6.** F

7. G   **8.** H   **9.** I

Graph and label each point on the coordinate plane at the right.

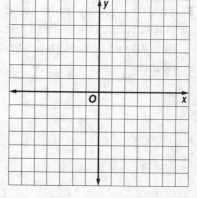

10. $J(2, 2)$ **11.** $K(-3, 4)$ **12.** $L(-4, 1)$

13. $M(-3, -3)$ **14.** $N(1, -4)$ **15.** $O(0, 0)$

16. $P(4, 5)$ **17.** $Q(4, -3)$ **18.** $R(-6, -5)$

19. ZOOS Use the map of the zoo at the right.
   a. What animals are located at $(-2, 3)$?

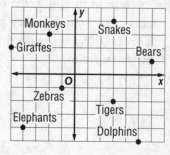

   b. In what quadrant are the bears located?

   c. Find the ordered pair that represents the location of the dolphins.

20. GEOMETRY Graph the points $A(-3, -1)$, $B(0, 4)$, $C(4, 3)$, and $D(1, -2)$ on the coordinate plane at the right. Connect the points from A to B, B to C, C to D, and D to A.

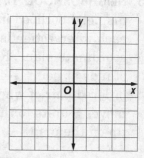

Get ConnectED *For more practice, go to www.connected.mcgraw-hill.com.*

Problem-Solving Practice

The Coordinate Plane

SCHOOL For Exercises 1–4, use the coordinate plane at the right. It shows a map of the rooms in a middle school.

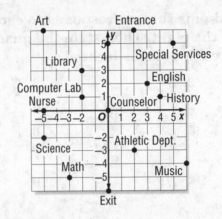

1. Thalia is in the room located at $(-2, 1)$. What room is she in? Describe in words how to get from the origin to this point.	**2.** Thalia's next class is 8 units to the right and 5 units down on the map from where she is now. In what room is Thalia's next class? Find the ordered pair that represents the location of that room.
3. Tyrone is in the Art room, but his next class is in the History room. Give Tyrone directions on how to get to the History room.	**4.** On the map, which classrooms are located in the third quadrant? Describe the coordinates of all points in the third quadrant.
5. NEIGHBORHOOD Ernie made a map of his neighborhood in such a way that each intersection is a point on a coordinate plane. Right now, Ernie stands at point $(-4, -3)$. Give the ordered pair of where he will be if he moves 5 units to the right and 7 units up on the map.	**6. NEIGHBORHOOD** Refer to Exercise 5. In which quadrant is Ernie when he is done walking? Describe this quadrant.

Course 2 • Integers

Multi-Part Lesson 2

PART B

Homework Practice

Add Integers

Add.

1. $34 + 22$

2. $-29 + 30$

3. $9 + (-32)$

4. $-16 + (-28)$

5. $4 + (-50)$

6. $-12 + (-63)$

7. $-42 + 42$

8. $-28 + 14$

9. $13 + 63$

10. $18 + (-12) + 5$

11. $-22 + (-10) + 15$

12. $-14 + 0 + 13$

Write an addition expression to describe each situation. Then find each sum and explain its meaning.

13. **WEIGHT** An actor gains 20 pounds for a part and then loses 15 pounds during the filming of the movie to go along with the story.

14. **TEMPERATURE** At 4:00 A.M., the outside temperature was $-28°$F. By 4:00 P.M. that same day, it rose 38 degrees.

ALGEBRA Evaluate each expression if $a = 12$, $b = -15$, and $c = -10$.

15. $a + (-12)$

16. $-20 + b$

17. $c + 23$

18. $b + c$

19. $a + c$

20. $a + b$

21. **ROLLER COASTERS** The latest thrill ride at a popular theme park takes roller coaster fans on an exciting ride. In the first 20 seconds, it carries its passengers up a 100-meter hill, plunges them down 72 meters, and quickly takes them back up a 48-meter rise. How much higher or lower from the start of the ride are they after these 20 seconds?

Get ConnectED *For more practice, go to* www.connected.mcgraw-hill.com.

Problem-Solving Practice

Add Integers

Write an addition expression to describe each situation. Then find each sum.

1. FOOTBALL A team gains 20 yards. Then they lose 7 yards.	**2. MONEY** Roger owes his mom $5. He borrows another $6 from her.
3. GOLF Jewel's score was 5 over par on the first 9 holes. Her score was 4 under par on the second 9 holes.	**4. HOT AIR BALLOON** A balloon rises 340 feet into the air. Then it descends 130 feet.
5. CYCLING A cyclist travels downhill for 125 feet. Then she travels up a hill 50 feet.	**6. AIRPLANE** A plane descends 1,200 feet. Then it descends another 500 feet.

Homework Practice

Subtract Integers

Subtract.

1. $16 - 14$

2. $-4 - 2$

3. $9 - (-2)$

4. $-6 - (-8)$

5. $-20 - 10$

6. $-28 - (-13)$

7. $-33 - 33$

8. $28 - 14$

9. $13 - (-63)$

10. $-18 - (-12)$

11. $52 - (-30)$

12. $-15 - 0$

13. **WEATHER** The highest and lowest recorded temperatures for the state of Texas are 120°F and −23°F. Find the difference in these extreme temperatures.

ALGEBRA Evaluate each expression if $x = -8$, $y = 7$, and $z = -11$.

14. $x - 7$

15. $-13 - y$

16. $-11 - z$

17. $x - z$

18. $z - y$

19. $y - x$

20. $x - (-z)$

21. $|y - z|$

22. $x - z - y$

23. **ANALYZE TABLES** In golf, scores are often stated as the number of strokes above or below par for the course. Four golfers played two rounds of golf during the weekend. The table lists their scores for each round in relation to par.

Golfer	Patrick	Diane	James	Judy
Round 1	−6	+1	+2	−3
Round 2	−2	−4	+7	+6

a. Find the difference between James's Round 2 score and Diane's Round 2 score.

b. Find the difference between Patrick's lower score and Judy's higher score.

Get ConnectED *For more practice, go to* www.connected.mcgraw-hill.com.

Multi-Part Lesson **2**
PART **D**

Problem-Solving Practice

Subtract Integers

---

1. FOOTBALL A team gained 5 yards on their first play of the game. Then they lost 6 yards. Find the total change in yardage.

2. CHECKING Your checking account is overdrawn by $50. You write a check for $20. What is the balance in your account?

---

3. TEMPERATURE The average temperature in Calgary, Canada, is 22°C in July and −11°C in January. Find the range of the highest and lowest temperatures in Calgary.

4. ROLLER COASTER A roller coaster begins at 90 feet above ground level. Then it descends 105 feet. Find the height of the coaster after the first descent.

---

5. SAVINGS Sonia has $235 in her savings account. She withdraws $45. What is left in her savings account?

6. BEACH Wai and Kuri were digging in the sand at the beach. Wai dug a hole that was 15 inches below the surface and Kuri dug a hole that was 9 inches below the surface. Find the difference in the depths of their holes.

Homework Practice

Problem-Solving Investigation: Look for a Pattern

Mixed Problem Solving

Use the *look for a pattern* strategy to solve Exercises 1 and 2.

1. **NUMBERS** What are the next two numbers in the pattern below?
3, 15, 75, 375, . . .

2. **QUILTING** Mrs. Perez is a talented quilter. In the center of the design of her quilt are four identical red squares in the shape of a square. Surrounding these red squares is a border of 12 identical white squares. Surrounding these white squares is a border of 20 identical blue squares. How many squares are in the next border surrounding the 20 blue squares?

B	B	B	B	B	B
B	W	W	W	W	B
B	W	R	R	W	B
B	W	R	R	W	B
B	W	W	W	W	B
B	B	B	B	B	B

3. The first 4 figures in a geometric pattern are shown.

How many dots would make up the 7th term of the pattern?

Use any strategy to solve Exercises 4–7.

4. **TRANSPORTATION** A college needs to transport the swim team to a state meet. The large van carries 15 people and each smaller van carries 9 people. How many smaller vans are needed to provide rides for 54 people if the large van is used?

5. **ALPHABET** What are the next two letters in each pattern shown?
D, H, L, P, ...

C, F, I, L, ...

6. **POPULATION** The land of Ebeye, an island in the Pacific, had a population of 200 in 1990, 230 in 1995, and 260 in 2000. About how many people will live there in 2015?

7. **ASTRONOMY** Earth is 93 million miles from the Sun, while Mars is 142 million miles from the Sun. Theoretically, what is the closest distance Mars could be to Earth?

Get ConnectED *For more practice, go to* www.connected.mcgraw-hill.com.

Multi-Part Lesson 3

PART A

Problem-Solving Practice

Problem-Solving Investigation: Look for a Pattern

Solve each problem using any strategy you have learned.

1. **COLLECTIONS** Brittany received 8 silver dollars on her eighth birthday. After her next birthday she had 15 and after the next she had 22. After her eleventh birthday she had 29 silver dollars. How many silver dollars will she have after her 16th birthday if her collection increases at the same rate every year?

2. **PATTERNS** List the next three terms in the following sequence.
27, 39, 51, 63, …

3. **GEOMETRY** There are 6 rows of squares stacked upon each other. The first three are shown. How many total squares are needed for the entire pattern?

4. **TICKET SALES** Madison High School is putting on a school play. They decide to charge $11 for main floor seats and $7 for balcony seats. If the school sold twice as many main floor seats as balcony seats and made $870, how many of each type of seat did they sell?

5. **EXERCISE** The table below shows the distance Katie ran each day this week. If Katie wants to run 30 miles a week, how many miles must she run on Sunday?

Monday	4 miles
Tuesday	7 miles
Wednesday	5 miles
Thursday	5 miles
Friday	2 miles
Saturday	3 miles

6. **AGE** Brad is three years more than half of Brandon's age. If their combined age is 93 years, how old is each man?

NAME _____ DATE _____ PERIOD _____

Homework Practice

Multiply Integers

Multiply.

1. $4(-7)$

2. $-14(5)$

3. $9(-12)$

4. $-6(-8)$

5. $27(-3)$

6. $-11(-13)$

7. $-55(0)$

8. $(-7)(-7)$

9. $78(-1)$

10. $(-3)^3$

11. $(-1)^4$

12. $(-8)^2$

13. Find -5 cubed.

14. Find the product of 13 and -31.

ALGEBRA Evaluate each expression if $a = -5$, $b = 4$, $c = -1$, and $d = 8$.

15. $5b$

16. $3c$

17. ad

18. $-7bd$

19. abc

20. $-5c^3$

21. $-a^2b$

22. $-4d - a$

23. $b^2 - 4ac$

24. RECREATION Hiking up a mountain, you notice that the air temperature drops 10°C for every 1,000 meters increase in elevation. Write a multiplication expression to represent the decrease in temperature if you hike up the mountain 3,000 meters. Then evaluate the expression and explain its meaning.

Get ConnectED *For more practice, go to* www.connected.mcgraw-hill.com.

Problem-Solving Practice

Multiply Integers

Multiply.

1. TEMPERATURE Suppose the temperature outside is dropping 3 degrees each hour. How much will the temperature change in 8 hours?	**2. DIVING** A deep-sea diver descends below the surface of the water at a rate of 60 feet each minute. What is the depth of the diver after 10 minutes?
3. STOCK A computer stock lost 2 points each hour for 6 hours. Describe the total change in the stock after 6 hours.	**4. DROUGHT** A drought can cause the level of the local water supply to drop by a few inches each week. Suppose the level of the water supply drops 2 inches each week. How much will it change in 4 weeks?
5. MONEY Mrs. Rockwell lost money on an investment at a rate of $4 per day. Describe the change in her investment after two weeks.	**6. TENNIS BALLS** Josh purchased 8 cans of tennis balls. The cans came with 3 balls in each can. How many tennis balls did Josh purchase?

Homework Practice

Divide Integers

Divide.

1. $42 \div (-7)$ **2.** $45 \div (-5)$ **3.** $-9 \div 3$

4. $-64 \div (-8)$ **5.** $-39 \div (-13)$ **6.** $-121 \div 11$

7. $\dfrac{-48}{12}$ **8.** $\dfrac{-35}{7}$ **9.** $\dfrac{-38}{-2}$

10. $\dfrac{32}{-16}$ **11.** $\dfrac{55}{-5}$ **12.** $\dfrac{-63}{7}$

13. Divide 75 by -25.   **14.** Find the quotient of -30 and -15.

ALGEBRA Evaluate each expression if $f = -15$, $g = 5$, and $h = -45$.

15. $-20 \div g$ **16.** $90 \div h$ **17.** $h \div f$

18. $fg \div 25$ **19.** $\dfrac{f - h}{10}$ **20.** $\dfrac{g - 5}{-1}$

21. $-f \div g$ **22.** $\dfrac{h - 3g}{f}$ **23.** $\dfrac{f + h}{-g}$

24. ZOOLOGY The table below shows the weight in pounds of large adult males in the cat family.

Cat	Cheetah	Cougar	Leopard	Lion	Tiger
Weight	143	227	200	550	400

a. What is the average weight of these cats?

b. What is the average weight of the two largest cats?

Get ConnectED *For more practice, go to* <u>www.connected.mcgraw-hill.com</u>.

Problem-Solving Practice

Divide Integers

Divide.

1. STOCK MARKET During a 5-day workweek, the stock market decreased by 65 points. Find the average daily change in the market for the week.	**2. MOTION** Mr. Diaz decreased the speed of his car by 30 miles per hour over a period of 10 seconds. Find the average change in speed each second.
3. WEATHER Over the past seven days, Mrs. Cho found that the temperature outside had dropped a total of 35 degrees. Find the average change in temperature each day.	**4. BASKETBALL** The basketball team lost their last 6 games. They lost by a total of 48 points. Find their average number of points relative to their opponents.
5. POPULATION The enrollment at Davis Middle School dropped by 60 students over a 5-year period. What is the average yearly drop in enrollment?	**6. SUBMARINE** A submarine descends at a rate of 60 feet each minute. How long will it take it to descend to a depth of 660 feet below the surface?

NAME _____ DATE _____ PERIOD _____

Homework Practice
Terminating and Repeating Decimals

Write each fraction or mixed number as a decimal. Use bar notation if the decimal is a repeating decimal.

1. $\dfrac{5}{8}$

2. $\dfrac{2}{9}$

3. $\dfrac{37}{16}$

4. $-\dfrac{3}{4}$

5. $\dfrac{27}{50}$

6. $\dfrac{121}{25}$

7. $-\dfrac{5}{6}$

8. $\dfrac{1}{33}$

9. $\dfrac{62}{11}$

10. $\dfrac{2}{3}$

11. $-\dfrac{11}{40}$

12. $\dfrac{13}{20}$

13. $\dfrac{83}{5}$

14. $\dfrac{3}{10}$

15. $-\dfrac{1}{9}$

16. $\dfrac{3}{7}$

17. $\dfrac{111}{24}$

18. $\dfrac{7}{32}$

Write each decimal as a fraction or mixed number in simplest form.

19. 0.4

20. -0.83

21. -3.75

22. 2.42

23. 0.16

24. 0.65

25. **KILOMETERS** One kilometer is approximately 0.62 mile. What fraction represents this length?

26. **MARATHON** Jake completed a marathon race in 3 hours and 12 minutes. Write Jake's running time as a decimal.

[icon] **Get ConnectED** *For more practice, go to* www.connected.mcgraw-hill.com.

Problem-Solving Practice

Terminating and Repeating Decimals

1. **BOYS AND GIRLS** There were 6 girls and 18 boys in Mrs. Johnson's math class. Write the number of girls as a fraction of the total class. Then write the fraction as a decimal.

2. **CATS** In a neighborhood of 72 families, 18 families own one or more cats. Write the number of families who own one or more cats as a fraction. Then write the fraction as a decimal.

3. **CELLULAR PHONES** In Italy, about 74 of every 100 people use cellular telephones. Write the fraction of cellular phone users in Italy. Then write the fraction as a decimal.

4. **FRUITS** Ms. Rockwell surveyed her class and found that 12 out of the 30 students chose peaches as their favorite fruit. Write the number of students who chose peaches as a fraction in simplest form. Then write the fraction as a decimal.

5. **TRAVEL** Tora took a short trip of 320 miles. He stopped to have lunch after he had driven 120 miles. Write the fraction of the trip he had completed by lunch in simplest form. Then write the fraction as a decimal.

6. **VOTING** In a recent school election, 208 of the 325 freshmen voted in their class election. Write the fraction of freshmen who voted. Then write the fraction as a decimal.

Multi-Part Lesson **1**

PART **C**

Homework Practice

Compare and Order Rational Numbers

Replace each ⬤ with >, <, or = to make a true sentence. Use a number line if necessary.

1. $\frac{5}{6}$ ⬤ $\frac{1}{3}$　　　　**2.** $\frac{4}{5}$ ⬤ $\frac{9}{10}$　　　　**3.** $\frac{6}{9}$ ⬤ $\frac{4}{6}$　　　　**4.** $\frac{2}{7}$ ⬤ $\frac{1}{8}$

5. $\frac{15}{21}$ ⬤ $\frac{12}{18}$　　　**6.** $\frac{24}{32}$ ⬤ $\frac{36}{48}$　　　**7.** $-\frac{8}{11}$ ⬤ $-\frac{10}{11}$　　**8.** $\frac{14}{15}$ ⬤ $\frac{19}{20}$

9. $4\frac{1}{5}$ ⬤ $4\frac{2}{10}$　　**10.** $7\frac{4}{9}$ ⬤ $7\frac{2}{3}$　　**11.** $-1\frac{17}{20}$ ⬤ $-1\frac{8}{10}$　　**12.** $9\frac{1}{2}$ ⬤ $9\frac{5}{6}$

13. 1 out of 2 ⬤ 8 out of 10　　　　　**14.** 0.65 ⬤ 65 out of 100

15. 4 out of 5 ⬤ $\frac{3}{4}$　　　　　　**16.** 1 out of 3 ⬤ 1.3

17. $\frac{2}{3}$ mile ⬤ $\frac{2}{5}$ mile　　　　　**18.** $\frac{7}{10}$ gram ⬤ 0.72 gram

19. $\frac{3}{8}$ yard ⬤ $\frac{1}{4}$ yard　　　　**20.** $2\frac{1}{2}$ quarts ⬤ $2\frac{3}{5}$ quarts

List each set of numbers in order from least to greatest.

21. $\frac{3}{5}, \frac{2}{3}$, 0.65　　　　**22.** $\frac{7}{8}$, 0.98, $\frac{8}{9}$　　　　**23.** 0.2, $\frac{1}{4}, \frac{1}{12}$

24. BASEBALL The pitchers for the home team had 12 strikeouts for 32 batters, while the pitchers for the visiting team had 15 strikeouts for 35 batters. Which pitching team had a greater fraction of strikeouts?

25. TRANSPORTATION To get to school, $\frac{19}{50}$ of the students ride in the family vehicle, 5 out of 12 students ride on the school bus, and 0.12 of the students ride a bike. Order the types of transportation students use to get to school from least to greatest.

Get ConnectED *For more practice, go to* www.connected.mcgraw-hill.com.

Multi-Part
Lesson **1**

PART **C**

Problem-Solving Practice

Compare and Order Rational Numbers

---

1. **RAIN** The amount of rainfall was measured after a recent storm. The north side of town received $\frac{7}{8}$ inch of rain, and the south side received $\frac{13}{15}$ inch of rain. Which side of town received more rain from the storm?

2. **MOVIES** Because he sees movies at his local theater so often, Delmar is being offered a discount. He can have either $\frac{1}{3}$ off his next ticket or $\frac{3}{10}$ off his next ticket. Which discount should Delmar choose? Explain.

---

3. **TRACK** Willie runs the 110-meter hurdles in $17\frac{3}{5}$ seconds, and Anier runs it in $17\frac{6}{11}$ seconds. Which runner is faster?

4. **FARMING** Cassie successfully harvested $\frac{7}{12}$ of her crop, and Robert successfully harvested $\frac{29}{50}$ of his crop. Which person successfully harvested the larger portion of his or her crop?

---

5. **TRANSPORTATION** My-Lien has enough room in her truck to move 3.385 tons of gravel. Her father has asked her to move $3\frac{5}{16}$ tons. Will My-Lien be able to move all of the gravel in only one trip? Explain.

6. **WOODWORKING** Kishi has a bolt that is $\frac{5}{8}$ inch wide, and she drilled a hole 0.6 inch wide. Is the hole large enough to fit the bolt? Explain.

---

7. **PIZZA** In a recent pizza-eating contest, Alfonso ate $1\frac{3}{8}$ pizzas, Della ate $1\frac{3}{10}$ pizzas, and Jack ate $1\frac{4}{9}$ pizzas. Which person won the contest?

8. **STUDYING** For a recent algebra exam, Pat studied $1\frac{8}{15}$ hours, Toni studied $1\frac{11}{20}$ hours, and Morgan studied $1\frac{9}{16}$ hours. List the students in order by who studied the most.

NAME _____ DATE _____ PERIOD _____

Homework Practice

Add and Subtract Like Fractions

Add or subtract. Write in simplest form.

1. $\dfrac{2}{5} + \dfrac{3}{5}$

2. $\dfrac{2}{9} + \dfrac{4}{9}$

3. $\dfrac{8}{11} - \dfrac{7}{11}$

4. $\dfrac{4}{8} + \dfrac{5}{8}$

5. $\dfrac{1}{18} + \dfrac{5}{18}$

6. $\dfrac{7}{15} - \dfrac{1}{15}$

7. $\dfrac{9}{16} - \dfrac{5}{16}$

8. $\dfrac{5}{14} - \dfrac{2}{14}$

9. $\dfrac{7}{8} - \dfrac{1}{8}$

10. $-\dfrac{7}{10} - \dfrac{4}{10}$

11. $\dfrac{5}{6} - \dfrac{3}{6}$

12. $-\dfrac{2}{3} - \left(-\dfrac{1}{3}\right)$

13. $\dfrac{5}{6} + \dfrac{1}{6}$

14. $\dfrac{5}{5} - \dfrac{3}{5}$

15. $\dfrac{4}{9} + \dfrac{8}{9}$

16. $\dfrac{5}{4} - \dfrac{1}{4}$

17. $\dfrac{2}{15} + \dfrac{4}{15} + \dfrac{1}{15}$

18. $\dfrac{7}{16} + \dfrac{1}{16} + \dfrac{3}{16}$

19. $\dfrac{3}{12} + \dfrac{1}{12} - \dfrac{11}{12}$

20. $\dfrac{4}{5} - \dfrac{7}{5} + \dfrac{1}{5}$

21. **STATES** Most of the state names in the United States end in a vowel ($a, e, i, o,$ or u). Of the 50 states, 25 of the state names end in either an a or an e and 6 end in either an i or an o. If none of the state names end in a u, what is the fraction of state names that end in a vowel?

22. **JIGSAW PUZZLES** Over the weekend, Halverson had put together $\dfrac{3}{16}$ of a jigsaw puzzle, while Jaime put together $\dfrac{10}{16}$ of the puzzle. Who had completed a greater fraction of the jigsaw puzzle, and by how much?

23. **TULIPS** Solan and Julie each planted tulips. Of Solan's 20 tulips, 15 were red, while 10 of Julie's 20 tulips were red. How much greater was Solan's fraction of red tulips than Julie's?

Get ConnectED *For more practice, go to* <u>www.connected.mcgraw-hill.com</u>.

Multi-Part Lesson 2
PART A

Problem-Solving Practice

Add and Subtract Like Fractions

RETAIL STORES For Exercises 1–4, use the table at the right. It shows what fraction of the stores at a mall fall into seven categories.

Type of Store	Fraction of Stores in Mall
jewelry	$\frac{1}{30}$
clothing	$\frac{16}{30}$
gifts	$\frac{5}{30}$
electronics	$\frac{1}{30}$
department	$\frac{2}{30}$
shoes	$\frac{2}{30}$
athletic	$\frac{3}{30}$

1. What fraction of the stores are jewelry or gift stores?

2. What fraction of the stores are clothing or electronics stores?

3. Which type of store has the greatest number of stores?

4. How many more clothing stores are there than athletic stores? Write as a fraction.

5. **SEWING** Jin wants to make a scarf and matching hat for his sister. The patterns call for $\frac{7}{8}$ yard of fabric for the scarf and $\frac{4}{8}$ yard of fabric for the hat. How much fabric should Jin buy?

6. **RESTAURANT** Ms. Malle owns a restaurant. Typically, $\frac{3}{20}$ of the customers order fish, while $\frac{7}{20}$ of the customers order poultry. What fraction of her customers order either fish or poultry?

Homework Practice

Add and Subtract Unlike Fractions

Add or subtract. Write in simplest form.

1. $\dfrac{1}{18} + \dfrac{5}{6}$

2. $\dfrac{7}{15} - \dfrac{1}{5}$

3. $\dfrac{9}{16} - \dfrac{5}{12}$

4. $\dfrac{5}{14} - \dfrac{2}{21}$

5. $\dfrac{7}{8} - \dfrac{1}{6}$

6. $-\dfrac{7}{10} - \dfrac{4}{15}$

7. $\dfrac{5}{6} - \left(-\dfrac{3}{4}\right)$

8. $-\dfrac{2}{3} - \left(-\dfrac{1}{2}\right)$

9. $1 + \dfrac{1}{6}$

10. $1 - \dfrac{3}{4}$

11. $4 + \dfrac{8}{9}$

12. $5 - \dfrac{1}{4}$

13. $\dfrac{2}{3} + \dfrac{4}{15} + \dfrac{1}{5}$

14. $\dfrac{3}{4} + \dfrac{1}{3} - \dfrac{11}{12}$

15. **EYE COLOR** If $\dfrac{2}{3}$ of the girls in class have brown eyes and $\dfrac{1}{4}$ of the girls have blue eyes, what fraction of the girls in class have neither blue or brown eyes?

16. **PIE** Ubi made a banana cream pie. His brother ate $\dfrac{1}{3}$ of the pie and his sister ate $\dfrac{2}{5}$ of the pie. How much less did his brother eat than his sister?

Get ConnectED *For more practice, go to* www.connected.mcgraw-hill.com.

Problem-Solving Practice

Add and Subtract Unlike Fractions

MARBLES For Exercises 1–4, use the table showing colors of marbles.

Color	Fraction
Red	$\frac{3}{50}$
Blue	$\frac{3}{25}$
Green	$\frac{3}{10}$
Yellow	$\frac{1}{25}$
Pink	$\frac{1}{10}$
Purple	$\frac{1}{5}$
White	$\frac{9}{50}$

1. What fraction of the marbles are red or blue?

2. What fraction of the marbles are green or purple?

3. What fraction represents how many more purple marbles there are than yellow ones?

4. What fraction represents how many more white marbles there are than pink ones?

5. GRADES If $\frac{1}{3}$ of the students got an A and $\frac{2}{5}$ of them got a B, what fraction of the students got an A or a B?

6. WATER AEROBICS If $\frac{5}{8}$ of the people in a water aerobics class are over age 65 and $\frac{1}{4}$ of the people in the class are under age 40, what fraction of the people in the class are either over 65 or under 40?

Multi-Part
Lesson 2
PART D

Homework Practice

Add and Subtract Mixed Numbers

Add or subtract. Write in simplest form.

1. $2\frac{3}{5} + 1\frac{4}{5}$

2. $3\frac{5}{6} - 1\frac{1}{6}$

3. $4\frac{3}{4} + 3\frac{1}{2}$

4. $6\frac{3}{8} - 2\frac{1}{4}$

5. $5\frac{9}{10} + 8\frac{2}{5}$

6. $3\frac{5}{8} - 2\frac{7}{8}$

7. $7\frac{5}{12} - 3\frac{3}{4}$

8. $1\frac{3}{5} + 2\frac{5}{6}$

9. $6 - 2\frac{3}{4}$

10. $3\frac{1}{2} + 2\frac{5}{8} - 4\frac{1}{4}$

11. GEOMETRY Find the perimeter of the triangle.

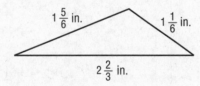

$1\frac{5}{6}$ in. $1\frac{1}{6}$ in.

$2\frac{2}{3}$ in.

12. KNITTING Nastia knitted two scarves for her dolls. One was $8\frac{3}{4}$ inches long. The other was $1\frac{1}{2}$ inches shorter than the first. How long was the second scarf?

Get ConnectED *For more practice, go to* www.connected.mcgraw-hill.com.

Multi-Part Lesson 2 PART D

Problem-Solving Practice

Add and Subtract Mixed Numbers

Solve. Write your answer as a fraction in simplest form.

1. **RUNNING** On Monday, Deborah ran $3\frac{2}{5}$ miles and on Tuesday she ran $4\frac{1}{5}$ miles. How many miles did she run on these two days together?

2. **PRINTING** Swamee and Luigi were printing calendars. Swamee used $2\frac{1}{2}$ ink cartridges while Luigi used $1\frac{3}{4}$ ink cartridges. How many more ink cartridges did Swamee use than Luigi?

3. **GARDENS** The table shows the number of pounds of green beans that Irma and Jeremiah each picked from their garden. How many total pounds of green beans did they pick?

Name	Pounds
Irma	$5\frac{2}{3}$
Jeremiah	$4\frac{5}{6}$

4. **MOVIES** Mr. and Mrs. Simpson went to two movies. The first movie lasted $2\frac{1}{3}$ hours and the second one lasted $1\frac{4}{5}$ hours. How much longer was the first movie?

5. **CELL PHONE** Mark talked on his cell phone 3 hours over the weekend. Genaro talked on his phone $1\frac{9}{10}$ hours. How much longer did Mark talk on his phone than Genaro?

6. **VACATION** Rodrick and Valentina drove to the coast. Rodrick drove $38\frac{9}{10}$ miles. Then Valentina drove the last $51\frac{3}{5}$ miles. How far did they drive to the coast?

Homework Practice

Multiply Fractions

Multiply. Write in simplest form.

1. $\frac{3}{5} \times \frac{1}{2}$

2. $\frac{3}{4} \times \frac{2}{7}$

3. $10 \times \frac{1}{3}$

4. $-\frac{5}{8} \times 7$

5. $\frac{1}{7} \times \frac{7}{9}$

6. $-\frac{6}{11} \times \left(-\frac{1}{6}\right)$

7. $\frac{5}{6} \times \frac{1}{5}$

8. $\frac{1}{8} \times \frac{4}{5}$

9. $\frac{3}{8} \times \frac{8}{9}$

10. $\frac{4}{7} \times \frac{21}{32}$

11. $-\frac{5}{8} \times \frac{18}{25}$

12. $\frac{20}{21} \times \frac{3}{5}$

13. $3\frac{1}{5} \times \frac{3}{8}$

14. $\frac{2}{3} \times \left(-4\frac{1}{3}\right)$

15. $15 \times 2\frac{2}{5}$

16. $5\frac{1}{2} \times 4$

17. $8 \times 3\frac{3}{8}$

18. $10 \times 1\frac{1}{15}$

19. $5\frac{1}{4} \times \left(-4\frac{2}{3}\right)$

20. $2\frac{2}{7} \times 1\frac{1}{8}$

For Exercises 21 and 22, use measurement conversions.

21. Find $\frac{1}{10}$ of $\frac{1}{100}$ of a meter.

22. Find $\frac{1}{60}$ of $\frac{1}{60}$ of an hour.

For Exercises 23–25, evaluate each verbal expression.

23. one fourth of two thirds

24. three fifths of one sixth

25. two fifths of one half

26. GASOLINE Jamal filled his gas tank and then used $\frac{7}{16}$ of the tank for traveling to visit his grandfather. He then used $\frac{1}{3}$ of the remaining gas in the tank to run errands around town. What fraction of the tank is filled with gasoline?

27. HIKING A hiker averages $6\frac{3}{8}$ kilometers per hour. If he hikes for $5\frac{1}{3}$ hours, how many kilometers does he hike?

ALGEBRA Evaluate each expression if $x = 3\frac{1}{3}$, $y = 4\frac{5}{6}$, and $z = 2$.

28. $x \times z - y$

29. $y \times z + x$

30. $3yz$

Get ConnectED *For more practice, go to* www.connected.mcgraw-hill.com.

Problem-Solving Practice

Multiply Fractions

---

1. POPULATION If $\frac{4}{5}$ of the population of a certain town is considered to be middle class and the population of the town is 2,000, how many people are considered middle class?

2. READING Robin has read $\frac{3}{4}$ of a book. Mark said he had read $\frac{1}{2}$ as much as Robin. What fraction of the book has Mark read?

---

3. RADIO A radio station spends $\frac{1}{40}$ of each 24 hours on public service announcements. How much time is spent on public service announcements each day?

4. SALE A bicycle is on sale for $\frac{2}{3}$ of its original price. If the original price is $354, what is the sale price?

---

5. STUDENT POPULATION One sixth of the students at a local college are seniors. The number of freshmen students is $2\frac{1}{2}$ times that amount. What fraction of the students are freshmen?

6. SEWING Anna wants to make 4 sets of curtains. Each set requires $5\frac{1}{8}$ yards of fabric. How much fabric does she need?

---

7. RUNNING It takes Awan $8\frac{1}{3}$ minutes to run one mile. It takes Kate $1\frac{1}{5}$ times longer. How long does it take Kate to run one mile?

8. STOCK Carl bought some stock at $25 a share. The stock increased to $1\frac{1}{2}$ times its value. How much is the stock per share?

Multi-Part
Lesson 3
c

Homework Practice

Problem-Solving Investigation: Draw a Diagram

Mixed Problem Solving

Use the *draw a diagram* strategy to solve Exercises 1 and 2.

1. **ANTS** An ant went 2 meters away from its nest searching for food. The next time, the ant went 3 meters away. Each successive time the ant leaves the nest to search for food, the ant travels the sum of the two previous times. How far will the ant travel on its fifth trip?

2. **NECKLACES** The center bead of a pearl necklace has a 16 millimeter diameter. Each successive bead in each direction is $\frac{3}{4}$ the diameter of the previous one. Find the diameter of the beads that are three away from the center bead.

Use any strategy to solve Exercises 3–6.

3. **TALENT SHOW** At a talent show, $\frac{3}{5}$ of the acts were singing. One-third of the remaining acts were instrumental. If 12 acts were instrumental, how many acts were in the talent show?

4. **GEOMETRY** Miss Greenwell is adding 4 feet to the length and width of her rectangular garden as shown in the diagram. How much additional area will the garden have?

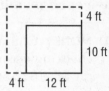

5. **YARD SALE** Myron has sold $18.50 worth of items at his yard sale. A neighbor bought two items and handed Myron a $10 bill. Myron returned $7.75 in change. How much has Myron now sold?

6. **COUNTRIES** The table shows the total land area of five countries.

Country	Total Area
Brazil	8.5 million sq km
Canada	10.0 million sq km
China	9.6 million sq km
Russia	17.1 million sq km
United States	9.6 million sq km

Estimate how much more total area Russia has than China.

Get ConnectED *For more practice, go to* www.connected.mcgraw-hill.com.

Multi-Part Lesson **3**

PART **C**

Problem-Solving Practice

Problem-Solving Investigation: Draw a Diagram

Solve each problem using any strategy you have learned.

---

1. MONEY Chantel has $125 left in her checking account after writing checks for $35, $22.50 and $16. What was her balance before she wrote the checks?

2. GEOMETRY Draw the next three figures in the pattern.

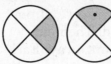

---

3. PIZZA Olivia has eaten $\frac{1}{3}$ of the pizza. If she has eaten 3 pieces, how many pieces were originally in the pizza?

4. EXERCISE Katlyn runs 2 miles after school each day, 3 miles on Saturday, and 4 miles on Sunday. How many miles does she run during one week?

---

5. WORK Jefferson wants to work at least 25 hours this week. If he has already worked 22 hours, how many hours does he need to work on Saturday?

6. TRAVEL The bus to Washington has traveled $\frac{5}{6}$ of the way there. If it has traveled 80 miles, how much farther does it have to go?

---

7. MUSEUMS The Art Club is planning on attending a museum. The admission cost is $10 for adults and $7.50 for students. If they plan on having 2 adults attend as chaperones and have $150 saved from a fundraiser, what is the maximum number of students who can attend?

8. SPORTS Janean made 50 baskets during the week at practice. The table below shows when she made the baskets. How many baskets did she make on Friday?

Day	Number of Baskets
Monday	5
Tuesday	12
Wednesday	16
Thursday	7
Friday	??

Multi-Part Lesson 3

PART D

Homework Practice

Divide Fractions

Divide. Write in simplest form.

1. $\dfrac{3}{5} \div \dfrac{3}{4}$

2. $-\dfrac{4}{7} \div \dfrac{8}{9}$

3. $\dfrac{6}{7} \div \dfrac{5}{6}$

4. $\dfrac{1}{4} \div \dfrac{1}{2}$

5. $7 \div \dfrac{1}{3}$

6. $\dfrac{6}{11} \div 2$

7. $4\dfrac{1}{5} \div (-7)$

8. $8 \div 4\dfrac{2}{3}$

9. $\dfrac{3}{4} \div 1\dfrac{1}{6}$

10. $-\dfrac{7}{9} \div \left(-2\dfrac{5}{8}\right)$

11. $3\dfrac{2}{5} \div 5\dfrac{1}{10}$

12. $4\dfrac{8}{9} \div \dfrac{2}{3}$

13. $2\dfrac{3}{5} \div 1\dfrac{1}{4}$

14. $7\dfrac{1}{2} \div 2\dfrac{1}{2}$

15. $5\dfrac{1}{4} \div \dfrac{7}{8}$

16. $-8\dfrac{1}{3} \div \dfrac{5}{9}$

17. **COOKING** Mrs. Lau rolls out $2\dfrac{3}{4}$ feet of dough to make noodles. If the noodles are $\dfrac{3}{8}$ of an inch wide, how many noodles will she make?

18. **PIZZA** Use the table that shows the weights of three sizes of pizza.

    a. How many times as heavy is the extra large pizza than the small pizza?

Pizza Size	Weight (lbs)
Extra large	$6\dfrac{1}{2}$
Medium	$3\dfrac{1}{4}$
Small	$1\dfrac{5}{8}$

    b. How many times as heavy is the medium pizza than the small pizza?

ALGEBRA Evaluate each expression if $a = \dfrac{2}{5}$, $b = \dfrac{3}{10}$, and $c = 2\dfrac{1}{2}$.

19. $b \div a$

20. $a \div c$

21. $3a \div b$

22. $\dfrac{1}{5}c \div a$

Get ConnectED *For more practice, go to* www.connected.mcgraw-hill.com.

Problem-Solving Practice

Divide Fractions

---

1. PUPPETS If a puppet requires $\frac{3}{4}$ yards of material, how many puppets can be made from 9 yards of material?

2. COOKING A batch of cookies requires $1\frac{1}{2}$ cups of sugar. How many batches can Ty make with $7\frac{1}{2}$ cups of sugar?

---

3. FOOD Julia has $3\frac{1}{2}$ pounds of dog food. She plans to split it equally among her 7 dogs. How much dog food will each dog receive?

4. SNOW CONES Randy has two 28-pound blocks of ice for his snow cone stand. If each snow cone requires $\frac{2}{3}$ pound of ice, how many snow cones can Randy make?

---

5. APPLES Juan took 6 apples and cut each into eighths. How many pieces of apple did he have?

6. VACATION The Torres family drove 1,375 miles during their $6\frac{1}{4}$-day vacation. Find the average number of miles they traveled each day.

---

7. RUNNING Hugo just joined the cross-country team and can run at a rate of $\frac{1}{7}$ mile each minute. How long will it take him to run a 5-mile race?

8. LUMBER Mrs. Shin has a piece of lumber that is $11\frac{5}{8}$ inches wide. She plans to split the width of lumber into 3 equal pieces. How wide will each piece be?

Homework Practice

Multiply and Divide Monomials

Simplify. Express using exponents.

1. $5^9 \cdot 5^4$

2. $3^8 \cdot 3^7$

3. $a \cdot a^4$

4. $m^5 \cdot m^2$

5. $3x \cdot 4x^6$

6. $-5d^4(6d^8)$

7. $(6k^5)(-k^5)$

8. $\left(\dfrac{3}{7}\right)^2\left(\dfrac{3}{7}\right)^3$

9. $(-4a^5)(6a^3)$

10. $(5a^5)(-3a^9)$

11. $\dfrac{5^{10}}{5^4}$

12. $\dfrac{8^3}{8}$

13. $\dfrac{b^6}{b^5}$

14. $\dfrac{G^{16}}{G^8}$

15. $\dfrac{18v^5}{3v^3}$

16. $\dfrac{24a^6}{6a^4}$

17. $\dfrac{30s^4}{-5s}$

18. $\dfrac{V^{30}}{V^{21}}$

19. $\dfrac{8n^{19}}{2n^{11}}$

20. $\dfrac{a^3 \cdot b^2}{a \cdot b^2}$

21. BONUSES A company has set aside 10^5 dollars for holiday employee bonuses. If the company has 10^3 employees and the money is divided equally among them, how much will each employee receive?

22. CAR LOANS After making a down payment, Mr. Green will make 7^2 monthly payments of 7^3 dollars each to pay for his new car. What is the total amount paid after the down payment?

🖥 **Get ConnectED** *For more practice, go to* <u>www.connected.mcgraw-hill.com.</u>

Multi-Part Lesson 4

PART A

Problem-Solving Practice

Multiply and Divide Monomials

1. **BOOKS** A publisher sells 10^5 copies of a new book. If each book sells for \$10, how much will the publisher make? Write your answer using exponents.

2. **TURKEYS** Mrs. Cowgill has 2^5 female turkeys on her farm. If each turkey lays 2^3 eggs, how many turkey eggs will she have? Write your answer using exponents.

3. **MONEY** 10^2 one-dollar bills are equivalent to 1 hundred-dollar bill. How many one-dollar bills are equivalent to 10^6 hundred-dollar bills? Write your answer using exponents.

4. **WEDDING** A couple is planning a meal for 3^5 people at their wedding. If they plan to seat 3^2 people at each table, how many tables do they need? Write your answer using exponents.

5. **COMPUTERS** The table shows the number of bytes in computer memory. How many times as great is a gigabyte than a kilobyte?

Memory Term	Number of Bytes
Byte	2^0 or 1
Kilobyte	2^{10}
Megabyte	2^{20}
Gigabyte	2^{30}

6. **SOUND** Sound is measured in decibels. Ordinary conversation is rated at about 60 decibels or a relative loudness of 10^6. A jet plane taking off is rated at about 110 decibels or a relative loudness of 10^{11}. How many times as great is the sound of a jet plane taking off than the sound of ordinary conversation?

Multi-Part Lesson 4

PART B

Homework Practice
Negative Exponents

Write each expression using a positive exponent.

1. 8^{-3}

2. $(-4)^{-5}$

3. $2k^{-4}$

4. $(-3)^{-3}$

5. 7^{-2}

6. $5a^{-3}$

Write each expression using a negative exponent other than –1.

7. $\dfrac{1}{9^3}$

8. $\dfrac{1}{32}$

9. $\dfrac{1}{b^7}$

10. $\dfrac{1}{m^4}$

11. $\dfrac{1}{100}$

12. $\dfrac{1}{12^2}$

Simplify each expression.

13. $a^{-5} \cdot a^3$

14. $6w^{-5} \cdot 8w^{-3}$

15. $\dfrac{f^{12}}{f^4}$

16. $4m^6 \cdot 3m^{-8}$

17. $\dfrac{81c^{-7}}{9c^{-5}}$

18. $\dfrac{w^{-9}}{w^{-5}}$

19. FRACTIONS Lorenzo needed to find $\frac{1}{9}$ of 27. His friend Carla told him to write $\frac{1}{9}$ as 3^{-2} and 27 as 3^3 and multiply 3^{-2} by 3^3 to get 3^{-6}. Was Carla correct? Explain.

20. RICE The mass of a grain of rice is about 10^{-2} grams. About how many grains of rice are in a container holding 10^4 grams of rice?

Get ConnectED *For more practice, go to* www.connected.mcgraw-hill.com.

Multi-Part Lesson 4

PART B

Problem-Solving Practice

Negative Exponents

1. **SALT** The mass of a grain of salt is about 10^{-4} gram. About how many grains of salt are in a shaker containing 10^3 grams?

2. **FISH** A certain fish swims at an average speed of 10^{-4} mile per minute. At this rate, how far would this fish travel in 30 minutes?

3. **SCIENCE** A virus has an average length of 10^{-7} meter. An atom has an average length of 10^{-10} meter. How many times is the average length of a virus than the average length of an atom?

4. **FOOTBALL** A football field is about 10^2 meters long. If there are 1,000 meters in one kilometer, how long is a football field in kilometers?

5. **SAND** A grain of sand has a volume of about 10^{-4} cubic millimeters. An empty bottle used to create sand art can hold about 10^{10} grains of sand. What is the approximate volume of the sand art bottle?

6. **SCIENCE** The pH of a substance is a measure of its acidity. The pH scale ranges from 0 to 14, with a pH of 7 being neutral. As the pH decreases, the substance is more acidic. For each increase of one in the pH level, the acidity of a substance is 10 times less. The pH of coffee is 5, and the pH of pure water is 7. Write a negative exponent that is equivalent to the ratio of the acidity level of coffee to the acidity level of pure water.

Multi-Part Lesson **4** PART **C**

Homework Practice
Scientific Notation

Express each number in standard form.

1. 6.42×10^2

2. 2.78×10^3

3. 2.357×10^5

4. 5.09×10^6

5. 3.6×10^{-2}

6. 5.1×10^{-5}

7. 9.82×10^{-4}

8. 3.42×10^{-3}

Express each number in scientific notation.

9. 356

10. 42,000

11. 8,350,000

12. 200,000

13. 0.11

14. 0.086

15. 0.000712

16. 0.0094

17. Which number is greater: 9.3×10^4 or 6.8×10^6?

18. Which number is less: 2.1×10^6 or 8.7×10^5?

19. POPULATION The table lists the populations of five countries. List the countries from least to greatest population.

Country	Population
Australia	2.1×10^7
Brazil	2.0×10^8
Egypt	8.3×10^7
Luxembourg	4.9×10^5
Singapore	4.7×10^6

20. SOLAR SYSTEM Saturn is 1.43×10^9 kilometers from the Sun. Write this number in standard form.

21. MEASUREMENT One cubic centimeter is about 0.061 cubic inch. Write this number in scientific notation.

22. DISASTERS In 1992, Hurricane Andrew caused over $25 billion in damage in Florida. Write $25 billion in scientific notation.

Get ConnectED *For more practice, go to* www.connected.mcgraw-hill.com.

Problem-Solving Practice

Scientific Notation

1. MEASUREMENT There is about 1.89×10^{-4} mile in a foot. Write this number in standard form.	**2. SPACE** The distance from Earth to Alpha Centauri is about 2.5×10^{13} miles. Write this number in standard form.
3. NATIONAL DEBT The national debt of the United States in 2008 was about $34,500 per person. Write this number in scientific notation.	**4. EARTH** The age of Earth is about 4.6×10^{9} years. Write this number in standard form.
5. PHYSICS The speed of sound is about 1.1×10^{3} feet per second. Write this number in standard form.	**6. ATOM** The mass of a sodium atom is about 0.0000000157 centimeter. Write this number in scientific notation.
7. CONCERT TOURS The table gives three top-grossing North American concert tours. Write the total gross for Band A's tour in scientific notation.	**8. MILITARY** In 2007, there were about 5.1×10^{5} U.S. Army personnel on active duty. Write this number in standard form.

Band	Total Gross (million $)
A	162
B	138.9
C	92.5

Homework Practice

Problem-Solving Investigation: Work Backward

Mixed Problem Solving

Use the *work backward* strategy to solve Exercises 1 and 2.

1. NUMBER THEORY A number is divided by 5. Then 3 is added to the quotient. After subtracting 10, the result is 30. What is the number?

2. COUPONS Kendra used 35 cents more in coupons at the store than Leanne. Leanne used 75 cents less than Becca, who used 50 cents more than Jaclyn. Jaclyn used 40 cents in coupons. What was the value of the coupons Kendra used?

Use any strategy to solve Exercises 3–6.

3. PATTERNS What are the next three numbers in the following pattern?

$$2, 3, 5, 9, 17, 33, \ldots$$

4. AGES Mr. Gilliam is 3 years younger than his wife. The sum of their ages is 95. How old is Mr. Gilliam?

5. GRAND CANYON The elevation of the North Rim of the Grand Canyon is 2,438 meters above sea level. The South Rim averages 304 meters lower than the North Rim. What is the average elevation of the South Rim?

6. WATER BILL The water company charges a residential customer $41 for the first 3,000 gallons of water used and $1 for every 200 gallons used over 3,000 gallons. If the water bill was $58, how many gallons of water were used?

Get ConnectED *For more practice, go to* www.connected.mcgraw-hill.com.

Problem-Solving Practice

Problem-Solving Investigation: Work Backward

For Exercises 1–3, use the information below.

WEATHER The temperature in Columbus, Ohio, on Monday is 35 degrees warmer than it was on Sunday. Saturday's temperature was 7 degrees cooler than Sunday's. At 45 degrees, Friday's temperature was 22 degrees warmer than Saturday's.

For Exercises 4–6, refer to the table below.

MONEY Shelly needs to go to the grocery store to get some items for a dinner party she is hosting with her brother Preston.

Green pepper	$1.79
Flank steak	$8.54
Wild rice	$3.29
Romaine lettuce	$3.79
Cucumber	$0.99

1. What was the temperature on Monday?

2. Estimate the average temperature for the time period from Saturday to Monday.

3. How many degrees cooler was the temperature on Friday than Monday?

4. How much money should she take to purchase the items contained in the table?

5. If Shelly has $24.00 in her purse before she goes to the store, how much will she have left after she shops?

6. If Preston pays Shelly for half the cost of the groceries, how much does he pay?

7. NUMBER THEORY How many different two-digit numbers can you make using the numbers 3, 7, 9, and 2 if no digit is repeated within a number?

8. PATTERNS The following numbers follow a pattern: 2, 8, 32, 128. What would the fifth number in the pattern be?

Course 2 • Equations and Inequalities

Homework Practice

Solve One-Step Addition and Subtraction Equations

Solve each equation. Check your solution.

1. $a + 4 = 11$ **2.** $6 = g + 8$ **3.** $x - 3 = -2$

4. $k + 8 = 3$ **5.** $j + 0 = 9$ **6.** $12 + y = 15$

7. $h - 4 = 0$ **8.** $m - 7 = 1$ **9.** $w + 5 = 4$

10. $b - 28 = 33$ **11.** $45 + f = 48$ **12.** $n + 7.1 = 8.6$

13. $-14 + t = 26$ **14.** $d - 3.03 = 2$ **15.** $10 = z + 15$

16. $c - 5.3 = -6.4$ **17.** $\frac{5}{12} + p = \frac{7}{12}$ **18.** $-\frac{1}{3} = -\frac{5}{6} + u$

For Exercises 19 and 20, write an equation. Use a bar diagram if needed. Then solve the equation.

19. CAFFEINE A cup of brewed tea has 54 milligrams less caffeine than a cup of brewed coffee. If a cup of tea has 66 milligrams of caffeine, how much caffeine is in a cup of coffee?

20. GEOMETRY The sum of the measures of the angles of a trapezoid is 360°. Find the missing measure.

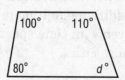

Get ConnectED *For more practice, go to* <u>www.connected.mcgraw-hill.com</u>.

Problem-Solving Practice

Solve One-Step Addition and Subtraction Equations

ANIMALS For Exercises 1–4, use the table.

The average lifespans of several different types of animals are shown in the table.

Average Lifespans of Animals			
Animal	**Lifespan (yr)**	**Animal**	**Lifespan (yr)**
Black bear	18	Guinea pig	4
Dog	12	Puma	?
Giraffe	10	Tiger	16
Gray squirrel	10	Zebra	?

1. The lifespan of a black bear is 3 years longer than the lifespan of a zebra. Write an addition equation that you could use to find the lifespan of a zebra.

2. Solve the equation you wrote in Exercise 1. What is the lifespan of a zebra?

3. The lifespan of a guinea pig is 8 years shorter than the lifespan of a puma. Write a subtraction equation that you could use to find the lifespan of a puma.

4. Solve the equation you wrote in Exercise 3. What is the lifespan of a puma?

5. TECHNOLOGY A survey of teens showed that teens in Pittsburgh aged 12–17 spend 15.8 hours per week online. Teens in Miami/Ft. Lauderdale spend 14.2 hours per week online. Write and solve an addition equation to find the difference in time spent online by teens in these cities.

6. SPORTS Annika Sorenstam won the 2006 MasterCard Classic with a final score of 8 under par, or −8. Her scores for the first two of the three rounds were −5 and −1. What was Ms. Sorenstam's score for the third round?

Homework Practice

Solve One-Step Multiplication and Division Equations

Solve each equation. Check your solution.

1. $\dfrac{k}{-11} = -3$

2. $16b = 32$

3. $72 = 12x$

4. $42 = 14y$

5. $\dfrac{x}{-16} = 1$

6. $-12k = -60$

7. $\dfrac{a}{13} = 0$

8. $-99 = 99y$

9. $\dfrac{h}{8} = 2$

10. $15 = \dfrac{y}{5}$

11. $\dfrac{h}{3} = -7$

12. $-1 = \dfrac{x}{-6}$

13. $9 = \dfrac{m}{2}$

14. $5b = -55$

15. $2z = 14$

16. $-3n = -45$

17. RAFFLE TICKETS Lavonne sold 4 times as many raffle tickets as Kenneth. Lavonne sold 56 raffle tickets. Write and solve an equation to find how many tickets Kenneth sold.

Get ConnectED *For more practice, go to* www.connected.mcgraw-hill.com.

Problem-Solving Practice

Solve One-Step Multiplication and Division Equations

For Exercises 1–8, write an equation. Then solve the equation.

1. EARNINGS Monica earned twice as much as Samuel mowing lawns. If Monica earned $48, how much did Samuel earn?

2. CHOIR The number of eighth graders in choir is three times the number of seventh graders. If there are 48 eight graders in choir, how many seventh graders are in choir?

3. CARS The cost of 6 motorcycles is equal to the cost of one SUV. If the SUV costs $30,000, find the cost of one motorcycle.

4. JUMP ROPES Carmen has a rope 54 feet long. She wants to cut it into 6-foot lengths to make jump ropes for the members of the jump roping team. How many jump ropes can Carmen make?

5. TAE KWON DO There are 8 competitors in each ring for a tae kwon do tournament. If there are 96 competitors in the tournament, how many rings do they need?

6. RAINFALL The amount of rainfall on Monday and Thursday is shown in the table. If the same amount of rain that fell on Monday fell for 3 days and the same amount that fell on Thursday fell for 2 days, how much rain would fall over those 5 days?

Day	Monday	Thursday
Rain (in.)	0.50	0.25

7. GERANIUMS Mary wants to put 4 geraniums in each pot. If she has 8 pots, how many geraniums should she buy?

8. HOMES The McClarens sold their house in Orlando, Florida, for $300,000. They split the income evenly among their four children. How much did each child get?

Multi-Part Lesson 2

PART D

Homework Practice

Solve Equations with Rational Coefficients

Find the multiplicative inverse or reciprocal of each number.

1. $\dfrac{7}{9}$

2. $\dfrac{5}{2}$

3. $\dfrac{1}{9}$

4. $\dfrac{1}{12}$

5. 4

6. 15

7. $4\dfrac{1}{3}$

8. $5\dfrac{4}{5}$

Solve each equation. Check your solution.

9. $-16 = 0.2b$

10. $12.3 = 0.41x$

11. $0.6h = 13.02$

12. $1 = \dfrac{x}{25}$

13. $0.9 = 0.4m$

14. $\dfrac{2}{3}t = 9$

15. $\dfrac{3}{7}g = 9$

16. $28 = \dfrac{4}{5}d$

17. $\dfrac{3}{8}n = \dfrac{1}{4}$

18. $\dfrac{2}{5} = \dfrac{4}{5}c$

19. $\dfrac{2}{3}z = 4\dfrac{1}{4}$

20. $\dfrac{5}{6}b = 1\dfrac{7}{8}$

21. $11.3y = 4.52$

22. $0.5y = 19.5$

23. $27.3 = \dfrac{3}{4}y$

24. $\dfrac{4}{7}x = -1.6$

25. DRAWING An architect needs to make a scale drawing of a home. The width w of the home in the drawing, in inches, is given by the equation $\dfrac{7}{8}w = 6$. What is the width of the home in the scale drawing?

26. VOLUNTEERS At a local shelter, 36 people volunteered to help prepare meals for disaster victims. If this represented $\dfrac{9}{16}$ of the volunteers at the shelter, write and solve an equation to determine how many volunteers helped at the local shelter.

Get ConnectED *For more practice, go to* www.connected.mcgraw-hill.com.

Multi-Part Lesson 2

PART D

Problem-Solving Practice

Solve Equations with Rational Coefficients

1. **BIKING** The speed s that Brent can ride his bike if he rides $\frac{3}{5}$ of an hour and travels 4 miles is given by the equation $4 = \frac{3}{5}s$. What is Brent's speed?

2. **BAND** The woodwind section of the middle school band makes up $\frac{1}{4}$ of the band. There are 9 members in the woodwind section. Use the equation $\frac{1}{4}m = 9$ to find the number of members m in the band.

3. **SALE** A coat is selling for $\frac{3}{4}$ of the original price. The sale price is $180. The original price p can be found using the equation $\frac{3}{4}p = 180$. Find the original price.

4. **SALARIES** Aaron's annual salary is $\frac{2}{3}$ as much as Dorie's salary. Aaron makes $46,000. Find Dorie's salary x using the equation $46,000 = \frac{2}{3}x$.

5. **ANIMALS** At a wildlife preserve, $\frac{1}{3}$ of the total number of reptiles and birds are reptiles. There are 14 reptiles. Use the equation $\frac{1}{3}a = 14$ to find the total number of reptiles and birds.

6. **SALES TAX** The sticker price p of a purchase with $\frac{1}{10}$ sales tax and a total price (including tax) of $5.28 can be found using the equation $\frac{11}{10}p = 5.28$. What is the sticker price?

7. **SEWING** Each costume uses $\frac{3}{4}$ yard of fabric. The number of costumes c that can be made using $11\frac{1}{4}$ yards of fabric can be found using the equation $\frac{3}{4}c = 11\frac{1}{4}$. Find the number of costumes that can be made.

8. **SAVINGS** Jasmine saves $46 each month from her part-time job. She saves $\frac{2}{5}$ of her earnings. Her earnings a can be found by using the equation $\frac{2}{5}a = 46$. Find her earnings.

Multi-Part Lesson 3

PART B

Homework Practice

Solve Two-Step Equations

Solve each equation. Check your solution.

1. $4h + 6 = 30$

2. $\frac{2}{7}y + 5 = -9$

3. $-3t + 6 = 0$

4. $-8 + 8g = 56$

5. $5k - 7 = -7$

6. $19 + 13x = 32$

7. $-\frac{1}{5}b - \frac{2}{5} = -2$

8. $-1n + 1 = 11$

9. $\frac{3}{4}f + 5 = -5$

10. $5d - 3.3 = 7.2$

11. $3 = 0.2m - 7$

12. $1.3z + 1.5 = 5.4$

13. KITTENS Kittens weigh about 100 grams when born and gain 7 to 15 grams per day. If a kitten weighed 100 grams at birth and gained 8 grams per day, in how many days will the kitten triple its weight?

14. TEMPERATURE Room temperature ranges from 20°C to 25°C. Find the range of room temperatures in °F. Use the formula $F - 32 = 1.8C$ to convert from the Celsius scale to the Fahrenheit scale.

Get ConnectED *For more practice, go to* www.connected.mcgraw-hill.com.

Course 2 • Equations and Inequalities

61

Multi-Part Lesson 3

PART B

Problem-Solving Practice

Solve Two-Step Equations

1. **GOLF** It costs $12 to attend a golf clinic with a local pro. Buckets of balls for practice during the clinic cost $3 each. How many buckets can you buy at the clinic if you have $30 to spend?

2. **MONEY** Paulo has $145 in his savings account. He earns $36 a week mowing lawns. If Paulo saves all of his earnings, after how many weeks will he have $433 saved?

3. **RETAIL** An online retailer charges $6.99 plus $0.55 per pound to ship electronics purchases. How many pounds is a DVD player for which the shipping charge is $11.94?

4. **MONEY** Caitlin has a $10 gift certificate to the music store. She has chosen a number of CDs from the $7 bargain bin. If the cost of the CDs is $32 after the gift certificate is credited, how many CDs did Caitlin buy?

5. **EMPLOYMENT** Mrs. Jackson earned a $500 bonus for signing a one-year contract to work as a nurse. Her salary is $22 per hour. If her first week's check including the bonus is $1,204, how many hours did Mrs. Jackson work?

6. **PHOTOGRAPHY** Alma subscribes to a website for processing her digital pictures. The subscription is $5.95 per month and 4-by-6-inch prints are $0.19 each. How many prints does Alma purchase if the charge for January is $15.83?

Course 2 • Equations and Inequalities

Homework Practice

Solve Equations with Variables on Each Side

Express each equation as another equivalent equation. Justify your answer.

1. $8x + 3 = 5x - 18$

2. $-7 - x = -11 - 2x$

3. $17 + 2x = 31 - 5x$

4. $6 - 3x = -18x$

Solve each equation. Check your solution.

5. $2.3 - 5x = 3x - 5.7$

6. $\frac{x}{3} + 4 = 5 + \frac{x}{6}$

7. $-9 + x = \frac{x}{8} - 23$

8. $-10 - 3x = 3x + 2$

9. $\frac{1}{3}x - 6 = 9 + \frac{2}{3}x$

10. $-x + 11 = -2.4 + x$

11. $5x + 3 = 4x$

12. $9 - 6x = 3x$

13. **MUGS** Manuela is having mugs made for a fundraiser. The Cup Company will make them for $4 each plus a $30 set-up charge. Mugs Are Us will make them for $4.50 each with no set-up charge. Write and solve an equation to find how many mugs Manuela can have made for the two company prices to be the same.

Problem-Solving Practice

Solve Equations with Variables on Each Side

1. PARTY Dwane was planning a graduation party. Caterer A charges $4 per person plus a delivery fee of $24. Caterer B has no delivery fee but charges $4.60 per person. How many people must Dwane invite to the party to make the cost of Caterer A the same as that of Caterer B?

2. CELL PHONES Use the table below that shows the rates for 3 cell phone companies.

Company	Per Minute	Per Month
Talk Now	$0.15	$20.00
Speedy Line	$0.00	$50.00
Phone Saver	$0.25	$5.00

How many minutes per month can you talk to make Talk Now cost the same as Phone Saver?

3. CELL PHONES Juan uses his cell phone about 100 minutes per month. Using the table in Exercise 2, which phone plan will be most economical for him?

4. CELL PHONES Shanequa uses her phone about 300 minutes per month. Which of the cell phone plans in Exercise 2 is the best plan for her?

5. NUMBER GAME Chase and Vijay were playing a game. Chase thought of a number. He said "30 more than 5 times my number is equal to 7 times my number." Vijay needed to guess the number. What number should Vijay guess?

6. BIKING Gladys rode her bike 20 miles per hour and stopped for 1 hour on the way to camp. Her brother Max left at the same time from the same place but biked at 24 miles per hour and stopped for $1\frac{1}{4}$ hours. Use the equation $\frac{d}{20} + 1 = \frac{d}{24} + 1\frac{1}{4}$ to find the distance they each traveled to camp.

Multi-Part Lesson 4

PART B

Homework Practice

Solve Inequalities by Addition or Subtraction

Solve each inequality.

1. $p + 9 < 7$

2. $t + 6 > -4$

3. $-12 \geq 7 + b$

4. $16 > -15 + k$

5. $25 < n - 11$

6. $-8 > h - 4$

7. $b - \dfrac{3}{4} < \dfrac{1}{2}$

8. $f - 5.2 \geq 1.6$

Solve each inequality. Graph the solution on a number line.

9. $n + 5 < 7$

10. $t + 2 > 10$

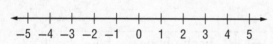

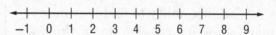

11. $p - 5 > -4$

12. $3 \leq \dfrac{1}{3} + n$

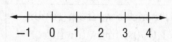

13. $4 \geq s - \dfrac{3}{4}$

14. $6.9 < w - 2.3$

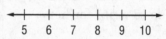

Write an inequality and solve each problem.

15. Four more than a number is no more than thirteen.

16. The difference of a number and −6 is less than 9.

17. Eleven less than a number is more than seventeen.

18. The sum of −8 and a number is at least 9.

19. **ENVELOPES** Sani has at least 68 envelopes to address. He has addressed 17 of them. Write and solve an inequality that describes how many more envelopes, at most, Sani has left to address.

Get ConnectED *For more practice, go to* www.connected.mcgraw-hill.com.

Course 2 • Equations and Inequalities

65

Multi-Part Lesson 4

PART B

Problem-Solving Practice

Solve Inequalities by Addition or Subtraction

1. DRIVING Louella is driving from Melbourne to Pensacola, a distance of more than 500 miles. After driving 240 miles, Louella stops for lunch. Write and solve an inequality to find how much farther Louella has to drive to reach Pensacola.

2. MONEY Aimee and Desmond are going to a play this evening. Desmond wants to have at least $50 in his wallet. He currently has $5. Write and solve an inequality to find how much more cash Desmond should put in his wallet.

3. FIELD TRIP There is space for 120 students to go on a field trip. Currently, 74 students have signed up. Write and solve an inequality to find how many more students can sign up for the field trip.

4. MUSIC Rogan is burning a music CD. The CD holds at most 70 minutes of music. Rogan has already selected 45 minutes of music. Write and solve an inequality to find how many more minutes of music Rogan can select.

5. HOMEWORK Petra must write a report with more than 1,000 words for her history class. So far, she has written 684 words. Write and solve an inequality to find how many more words Petra needs to write for her report.

6. HEIGHT Leslie hopes to be at least 72 inches tall. Right now he is 56 inches tall. Write an solve an inequality to find how much more Leslie would like to grow.

7. INTERNET Julius is allowed to surf the Internet for only 3 hours a week. He has already been online for $1\frac{2}{3}$ hours this week. Write and solve an inequality to find how much more time Julius can spend online this week.

8. GROCERIES The table shows how much Colleen has spent at the grocery store this week. To stay within her budget, she can spend only $90 per week on groceries. Write and solve an inequality to find how much more Colleen can spend at the grocery store this week.

Day	Amount Spent ($)
Monday	28
Wednesday	39

Multi-Part Lesson 4

PART C

Homework Practice

Solve Inequalities by Multiplication or Division

Solve each inequality. Graph the solution set on a number line.

1. $-8 \leq 8w$

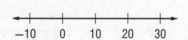

2. $-6a > 66$

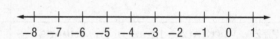

3. $-25t \leq -500$

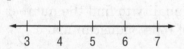

4. $18 > -3g$

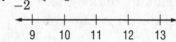

5. $\dfrac{y}{4} \leq 1.6$

6. $\dfrac{r}{-2} < -6$

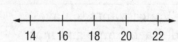

7. $-8 > \dfrac{k}{-0.2}$

8. $\dfrac{m}{-8} \leq -2.4$

Solve each inequality. Check your solution.

9. $13a + 26 \geq -13$

10. $-24 \leq 3b - 9$

11. $-3m - 6 \geq -27$

12. $-11 \geq \dfrac{c}{2} - 3$

13. $1 > \dfrac{y}{0.2} + 20$

14. $\dfrac{-1}{3}x + 3 > -1$

15. TEXT MESSAGES Nadine can send or receive a text message for $0.15 or get an unlimited number for $5.00. Write and solve an inequality to find how many messages she can send and receive so the unlimited plan is cheaper than paying for each message.

Get ConnectED *For more practice, go to* www.connected.mcgraw-hill.com.

Problem-Solving Practice

Solve Inequalities by Multiplication or Division

1. **PLANTS** Trini needs more than 51 cubic feet of soil to top up his raised garden. Each bag of soil contains 1.5 cubic feet. Write and solve an inequality to find how many bags of soil Trini needs.

2. **PETS** Becky wants to buy some fish for her aquarium. She has $20 to spend and the fish cost $2.50 each. Write and solve an inequality to find how many fish Becky can afford.

3. **PIZZA** Vikram and four of his friends are planning to split a pizza. They want to spend at most $4 per person. Write and solve an inequality to find the maximum cost of the pizza they can order.

4. **ROLLS** Sadie wants to make several batches of rolls. She has 13 tablespoons of yeast left in the jar and each batch of rolls takes $3\frac{1}{4}$ tablespoons. Write and solve an inequality to find the number of batches of rolls Sadie can make.

5. **CONSTRUCTION** Vance wants to have pictures framed. Each frame and mat costs $32 and he has at most $150 to spend. Write and solve an inequality to find the number of pictures he can have framed.

6. **RECTANGLE** You are asked to draw a rectangle with a width of 5 inches and an area less than 55 square inches. Write and solve an inequality to find the length of the rectangle.

7. **BABYSITTING** Hermes gets $4 an hour for babysitting. He needs to earn at least $100 for a stereo. Write and solve an inequality to find the number of hours he must babysit to earn enough for the stereo.

8. **TIME** The table shows how many minutes per day Terri spends on the phone and watching television. If she has 180 minutes in the day for leisure activities, write and solve an inequality to find the number of minutes she can spend listening to music.

Activity	Number of Minutes
Talking on phone	25
Watching television	120

Course 2 • Equations and Inequalities

Multi-Part Lesson 1

PART B

Homework Practice

Rates

Find each unit rate. Round to the nearest hundredth if necessary.

1. $11.49 for 3 packages

2. 2,550 gallons in 30 days

3. 88 students for 4 classes

4. 15.6°F in 13 minutes

5. 175 Calories in 12 ounces

6. 258.5 miles in 5.5 hours

7. 549 vehicles on 9 acres

8. $920 for 40 hours

9. 13 apples for 2 pies

10. MANUFACTURING A machinist can produce 114 parts in 6 minutes. At this rate, how many parts can the machinist produce in 15 minutes?

11. RECIPES A recipe that makes 8 jumbo blueberry muffins calls for $1\frac{1}{2}$ teaspoons of baking powder. How much baking powder is needed to make 3 dozen jumbo muffins?

Estimate the unit price for each item. Justify your answers.

12. $299 for 4 tires

13. 3 yards of fabric for $13.47

14. UTILITIES Use the table that shows the average monthly electricity and water usage.

   a. Which family uses about twice the amount of electricity per person than the other two families? Explain your reasoning.

Family Name	Family Size	Electricity (kilowatt-hours)	Water (gal)
Melendez	4	1,560	3,500
Barton	6	2,130	6,400
Stiles	2	1,490	2,500

   b. Which family uses the least amount of water per person? Explain your reasoning.

Get ConnectED *For more practice, go to www.connected.mcgraw-hill.com.*

Problem-Solving Practice

Rates

1. **TRAVEL** During Tracy's trip across the country, she traveled 2,884 miles. Her trip took 7 days. Find a unit rate to represent the average miles she traveled per day during the trip.

2. **BUDGET** Steve was offered $5,025 per year for a weekend lifeguarding job at a local pool. He wants to know how much his monthly income will be at this salary level. What is his rate of pay in dollars per month?

3. **MUSIC** Randall recorded 8 songs on his most recent CD. The total length of the CD is 49 minutes. Find a unit rate to represent the average length per song on the CD.

4. **CARPETING** Hana paid $1,200 for the carpet in her living room. The room has an area of 251.2 square feet. What was her unit cost of carpeting in dollars per square foot? Round to the nearest cent.

5. **SHOPPING** An 8-ounce can of tomatoes costs $1.14. A 12-ounce can costs $1.75. Which can of tomatoes costs the least per ounce?

6. **PETS** Last month, Hao's dog ate 40 cans of dog food in 30 days. How many cans should Hao buy to feed his dog for the next 6 days?

Homework Practice

Proportional and Nonproportional Relationships

1. **ANIMALS** The world's fastest fish, a sailfish, swims at a rate of 69 miles per hour. Is the distance a sailfish swims proportional to the number of hours it swims?

FOSSILS Use the following information.

In July, a paleontologist found 368 fossils at a dig. In August, she found about 14 fossils per day.

2. Is the number of fossils the paleontologist found in August proportional to the number of days she spent looking for fossils that month?

3. Is the total number of fossils found during July and August proportional to the number of days the paleontologist spent looking for fossils in August?

Get ConnectED *For more practice, go to* www.connected.mcgraw-hill.com.

NAME _____ DATE _____ PERIOD _____

Problem-Solving Practice

Proportional and Nonproportional Relationships

Use a table of values when appropriate to explain your reasoning.

1. SPORTS A touchdown is worth 6 points. Additionally you score an extra point if you can kick a field goal. Is the total number of points scored proportional to the number of touchdowns?	**2. RECREATION** An outdoor swimming pool costs $8 per day to visit during the summer. There is also a $25 yearly registration fee. Is the total cost proportional to the total number of days visited?
3. SCHOOL At a certain middle school, there are 26 students per teacher in every homeroom. Is the total number of students proportional to the number of teachers?	**4. TEAMS** A baseball club has 18 players for every team, with the exception of four teams that have 19 players each. Is the number of players proportional to the number of teams?
5. MONEY At the beginning of the summer, Conan had $180 in the bank. Each week he deposits another $64 that he earns mowing lawns. Is his account balance proportional to the number of weeks since he started mowing lawns?	**6. SHELVES** A bookshelf holds 43 books on each shelf. Is the total number of books proportional to the number of shelves in the bookshelf?

Homework Practice

Solve Proportions

Solve each proportion.

1. $\dfrac{b}{5} = \dfrac{8}{16}$

2. $\dfrac{18}{x} = \dfrac{6}{10}$

3. $\dfrac{t}{6} = \dfrac{30}{36}$

4. $\dfrac{11}{10} = \dfrac{n}{14}$

5. $\dfrac{2.5}{35} = \dfrac{2}{d}$

6. $\dfrac{3.5}{18} = \dfrac{z}{36}$

7. $\dfrac{0.45}{4.2} = \dfrac{p}{14}$

8. $\dfrac{2.4}{6} = \dfrac{2.8}{s}$

9. $\dfrac{3.6}{k} = \dfrac{0.2}{0.5}$

For Exercises 10–12, assume all situations are proportional.

10. **CLASSES** For every girl taking classes at the martial arts school, there are 3 boys who are taking classes at the school. If there are 236 students taking classes, write and solve a proportion to predict the number of boys taking classes at the school.

11. **BICYCLES** An assembly line worker at Rob's Bicycle factory adds a seat to a bicycle at a rate of 2 seats in 11 minutes. Write a proportion relating the number of seats s to the number of minutes m. At this rate, how long will it take to add 16 seats? 19 seats?

12. **PAINTING** Lisa is painting a fence that is 26 feet long and 7 feet tall. A gallon of paint will cover 350 square feet. Write and solve a proportion to determine how many gallons of paint Lisa will need.

Get ConnectED *For more practice, go to* www.connected.mcgraw-hill.com.

Multi-Part Lesson 1

PART D

Problem-Solving Practice

Solve Proportions

1. **USAGE** A 12-ounce bottle of shampoo lasts Enrique 16 weeks. How long would you expect an 18-ounce bottle of the same brand to last him?

2. **COMPUTERS** About 13 out of 20 homes have a personal computer. On a street with 60 homes, how many would you expect to have a personal computer?

3. **SNACKS** A 6-ounce package of fruit snacks contains 45 pieces. How many pieces would you expect in a 10-ounce package?

4. **TYPING** Ingrid types 3 pages in the same amount of time that Tanya types 4.5 pages. If Ingrid and Tanya start typing at the same time, how many pages will Tanya have typed when Ingrid has typed 11 pages?

5. **SCHOOL** A grading machine can grade 48 multiple-choice tests in 1 minute. How long will it take the machine to grade 300 tests?

6. **AMUSEMENT PARKS** The waiting time to ride a roller coaster is 20 minutes when 150 people are in line. How long is the waiting time when 240 people are in line?

7. **PRODUCTION** A shop produces 39 wet suits every 2 weeks. How long will it take the shop to produce 429 wet suits?

8. **FISH** Of the 50 fish that Alan caught from the lake, 14 were trout. The estimated population of the lake is 7,500 fish. About how many trout would you expect to be in the lake?

Multi-Part Lesson 2

PART A

Homework Practice

Problem-Solving Investigation: Draw a Diagram

Mixed Problem Solving

Use the *draw a diagram* strategy to solve Exercises 1 and 2.

1. **SWIMMING** Jon is separating the width of the swimming pool into equal-sized lanes with rope. It took him 30 minutes to create 6 equal-sized lanes. How long would it take him to create 4 equal-sized lanes in a similar swimming pool?

2. **TRAVEL** Two planes are flying from San Francisco to Chicago, a distance of 1,800 miles. They leave San Francisco at the same time. After 30 minutes, one plane has traveled 25 more miles than the other plane. How much longer will it take the slower plane to get to Chicago than the faster plane if the faster plane is traveling at 500 miles per hour?

Use any strategy to solve Exercises 3–6.

3. **TALENT SHOW** In a solo singing and piano playing show, 18 people sang and 14 played piano. Six people both sang and played piano. How many people were in the singing and piano playing show?

4. **LETTERS** Suppose you have three strips of paper as shown. How many capital letters of the alphabet could you form using one or more of these three strips for each letter? Assume the strips cannot be bent or creased. List them according to the number of strips.

5. **CLOTHING** A store has 255 wool ponchos to sell. There are 112 adult-sized ponchos that sell for $45 each. The rest are kid-sized and sell for $32 each. If the store sells all the ponchos, how much money will the store receive?

6. **DINOSAURS** Al made a model of a *Stegosaurus*. If you multiply the model's length by 8 and subtract 4, you will find the length of an actual *Stegosaurus*. If the actual *Stegosaurus* is 30 feet long, how long is Al's model?

Get ConnectED *For more practice, go to www.connected.mcgraw-hill.com.*

Problem-Solving Practice

Problem-Solving Investigation: Draw a Diagram

Use the *draw a diagram* strategy to solve each problem.

1. **TILING** Ruth is using 3-inch square tiles to cover a 4-foot by 2-foot area. The tiles are 0.5 inches tall. If the tiles were stacked on top of each other to create a tower, how many inches tall would the tower be?

2. **AQUARIUM** An aquarium holds 42 gallons of water. After 2 minutes, the aquarium has 3 gallons of water in it. How many more minutes will it take to completely fill the aquarium?

3. **FABRIC** It takes Lucy 7 minutes to cut a 20-yard by 1-yard roll of fabric into 14 equal pieces. How many minutes would it take her to cut a 30-yard by 1-yard roll of fabric into 25 equal pieces?

4. **SPORTS** The width of a soccer field is 12 feet more than $\frac{2}{3}$ of its length. If the field is 96 feet long, what is its perimeter?

5. **BEVERAGES** It requires 4 gallon jugs of water to fill 104 glasses equally. How many gallon jugs are required to fill 338 glasses equally?

6. **GAS** It takes Richard 48 seconds to fill his gas tank with 3 gallons of gas. If the tank holds 14 gallons, how many more seconds will it take to fill it completely?

Homework Practice

Scale Drawings

Use the diagram of a section of
the art museum shown. Use a
ruler to measure.

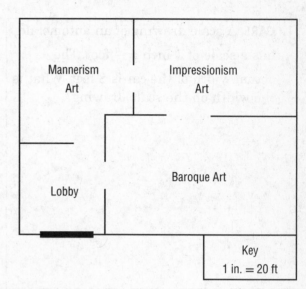

1. What is the actual length of the
   Impressionism Art room?

2. Find the actual dimensions of the *Baroque
   Art* room.

3. Find the scale factor for this blueprint.

Key
1 in. = 20 ft

Find the length of each model. Then find the scale factor.

4.
|←——60 ft——→|

1 in. = 8 ft

5.
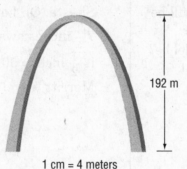

192 m

1 cm = 4 meters

6.
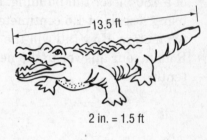
|←——13.5 ft——→|

2 in. = 1.5 ft

7. **SKYSCRAPER** A model of a skyscraper is made using a scale of
   1 inch:75 feet. What is the height of the actual building if the
   height of the model is $19\frac{2}{5}$ inches?

8. **GEOGRAPHY** Salem and Eugene, Oregon, are 64 miles apart.
   If the distance on the map is $3\frac{1}{4}$ inches, find the scale of the map.

9. **PYRAMIDS** The length of a side of the Great Pyramid of Khufu at
   Giza, Egypt, is 751 feet. If you were to make a model of the pyramid
   to display on your desk, which would be an appropriate scale: 1 in. = 10 ft
   or 1 ft = 500 ft? Explain your reasoning.

Get ConnectED *For more practice, go to* www.connected.mcgraw-hill.com.

Problem-Solving Practice

Scale Drawings

---

1. CARS A scale drawing of an automobile has a scale of 1 inch = $\frac{1}{2}$ foot. The actual width of the car is 8 feet. What is the width on the scale drawing?

2. MODELS A model ship is built to a scale of 1 centimeter : 5 meters. The length of the model is 30 centimeters. What is the actual length of the ship?

---

3. BUILDING Curtis wants to build a model of a 180-meter tall building. He will be using a scale of 1.5 centimeters = 3.5 meters. How tall will the model be? Round your answer to the nearest tenth.

4. TRAVEL Merritt is driving to Mount Shasta. On her map, she is a distance of $7\frac{3}{4}$ inches away. The scale of the map is $\frac{1}{2}$ inch = 50 miles. How far must Merritt travel to reach her destination?

---

5. MAPS A map of Levi's property is being made with a scale of 2 centimeters: 3 meters. What is the scale factor?

6. LANDSCAPING A pond is being dug according to plans that have a scale of 1 inch = 6.5 feet. The maximum distance across the pond is 9.75 inches on the plans. What will be the actual maximum distance across the pond?

Multi-Part Lesson 3

PART A

Homework Practice

Similar Figures

1. Which rectangle is similar to rectangle *RSTU*?

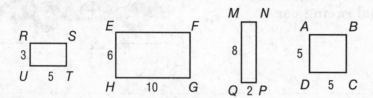

2. Which triangle is similar to triangle *XYZ*?

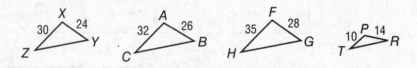

Find the value of *x* in each pair of similar figures.

3.

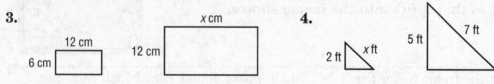

4.

5.

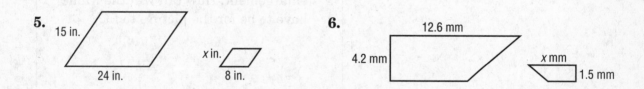

6.

7. FLAGPOLES Tasha wants to find the height of the flagpole at school. One morning, she determines the flagpole casts a shadow of 12 feet. If Tasha is 5 feet tall and casts a shadow of 3 feet, what is the height of the flagpole?

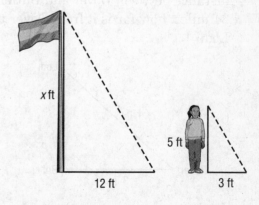

Get ConnectED *For more practice, go to* www.connected.mcgraw-hill.com.

NAME _____ DATE _____ PERIOD _____

Problem-Solving Practice

Similar Figures

MODEL CARS Use the following information. A scale model racing car is 11 inches long, 3 inches wide, and 2 inches tall. The actual racing car is shown at the right.

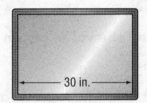
22 ft

1. How wide is the actual racing car?	**2.** How tall is the actual racing car?

PHOTOGRAPHY Use the given information. James wants to enlarge a photograph that is 6 inches wide and 4 inches tall so that it fits into the frame shown.

30 in.

3. How tall must the frame be for the picture to fit?

4. Suppose James cuts 1 inch from the width of the photo, so that it is 5 inches wide, before he makes the enlargement. How tall will the frame have to be for the picture to fit?

5. MAPS The map below shows the towns of Dover, Butler, and Lodi. If the actual distance between Dover and Butler is 24 miles, how far is it from Dover to Lodi?

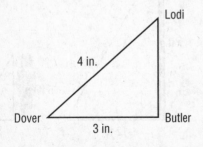
Lodi
4 in.
Dover
3 in.
Butler

6. BLUEPRINTS A blueprint for a house is shown below. If the front of the house is actually 30 feet wide, how tall is the house?

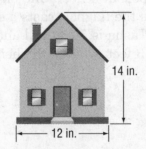
14 in.
12 in.

Multi-Part Lesson **3**

PART **B**

Homework Practice

Perimeter and Area of Similar Figures

For each pair of similar figures, find the perimeter of the second figure.

1.

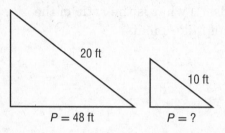

20 ft

10 ft

P = 48 ft   P = ?

2.

2 in.   3 in.

P = 12 in.   P = ?

3.

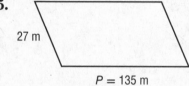

10 cm

5 cm

P = 30 cm   P = ?

4.

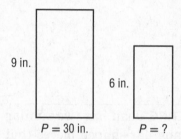

9 in.   6 in.

P = 30 in.   P = ?

5.

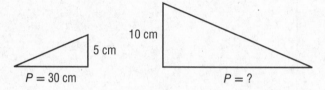

27 m   18 m

P = 135 m   P = ?

6. A triangle has a side length of 4 inches and an area of 18 square inches and a larger similar triangle has a corresponding side length of 8 inches. Find the area of the larger triangle.

7. A rectangle has a side length of 3 feet and an area of 24 square feet. A larger similar rectangle has a corresponding side length of 9 feet. Find the area of the larger rectangle.

8. FLOWER GARDEN A rectangular shaped flower garden has a length of 5 yards and an area of 15 square yards. A neighbor's flower garden is similar and has a length of 7 yards. What is the area of the neighbor's flower garden? Round your answer to the nearest whole number.

 **Get ConnectED** *For more practice, go to* <u>www.connected.mcgraw-hill.com.</u>

Problem-Solving Practice

Perimeter and Area of Similar Figures

1. **MATS** Mike has two mats that are in the shape of triangles. The scale factor of the two triangular mats is $\frac{7}{9}$. What is the ratio of the perimeters?

2. **AREA** Using the same mats as in Exercise 1, what is the ratio of the areas of Mike's mats?

3. **QUILTING** Joan has two similar rectangular-shaped pieces that she is using for her quilt. One has width 4 inches and length 6 inches and the other has a width of 6 inches. What is the perimeter of the larger rectangular piece?

4. **AREA** Using the same quilting pieces from Exercise 3, what is the area of the larger rectangular piece?

5. **GEOMETRY** Draw similar rectangles that are in the ratio of $\frac{2}{3}$. What is the ratio of the areas of the rectangles you drew?

6. **GEOMETRY** Draw similar triangles whose sides are in the ratio of $\frac{3}{4}$. What is the ratio of the perimeters of the triangles you drew?

Homework Practice
Percent of a Number

Find each number. Round to the nearest hundredth if necessary.

1. 55% of 140

2. 40% of 123

3. 37% of $150

4. 25% of 96

5. 11% of $333

6. 99% of 14

7. 140% of 30

8. 165% of 10

9. 150% of 150

10. 225% of 16

11. 106% of $40

12. 126% of 350

13. 4.1% of 30

14. 8.9% of 75

15. 24.2% of $120

16. SALES Mr. Redding sells vehicles to 20% of the people that come to the sales lot. If 65 people came last month, how many vehicles did he sell?

Find each number. Round to the nearest hundredth if necessary.

17. $\frac{5}{6}$% of 600

18. $30\frac{1}{3}$% of 3

19. 1,000% of 87

20. 100% of 56

21. 0.25% of 150

22. 0.7% of 50

23. ANALYZE TABLES Use the table that shows the percents of blood types of 145 donors during a recent blood drive.

a. Find how many donors had type B blood. Round to the nearest whole number if necessary.

b. How many donors did *not* have type O blood? Round to the nearest whole number if necessary.

c. Which blood type(s) had less than 10 donors?

Blood Type	Percent
O	45
A	40
B	11
AB	4

Problem-Solving Practice

Percent of a Number

SPORTS For Exercises 1 and 2, use the graph below. It shows the results of a poll of 440 ninth-grade students. Round answers to the nearest whole number.

PETS For Exercises 3 and 4, use the table below. It shows the pet ownership in a town of 1,650 households. Round answers to the nearest whole number.

Favorite Sports of Students

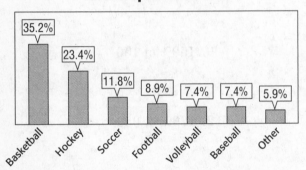

Pets in Household	Percent
At least one dog or cat	26.7
At least one dog	19.9
At least one cat	13
At least one dog and one cat	6.19

1. Write the percent as a fraction to find how many students surveyed chose hockey as their favorite sport. Solve.	**2.** How many students surveyed chose basketball as their favorite sport?
3. Write the percent as a decimal to find how many households have at least one dog. Solve.	**4.** How many households have at least one dog and one cat?
5. VOTING Going into a recent election, only about 62% of people old enough to vote were registered. In a community of about 55,200 eligible voters, how many people were registered?	**6. COLLEGE** A local college recently reported that enrollment increased to 108% percent of last year. If enrollment last year was at 17,113, about how many students enrolled this year? Round to the nearest whole number.

Homework Practice

Percent and Estimation

Estimate.

1. 39% of 80

2. 31% of 40

3. 28% of 110

4. 74% of 160

5. 87% of 19

6. 91% of 82

7. 34% of 59

8. 66% of 148

9. 9% of 71

10. 73% of 241

11. 126% of 80

12. 234% of 145

13. $\frac{1}{3}$% of 307

14. $\frac{1}{4}$% of 798

15. 1.1% of 62

16. 4.1% of 101

17. 67% of 11.9

18. 31% of 68.7

19. 9.8% of 359

20. 97.9% of 39

21. 52% of 57.9

22. 33% of 15.3

23. 21.1% of 151

24. 2.9% of 61.2

25. **ELEVATION** The highest point in Arizona is Humphreys Peak, with an elevation of 12,633 feet. Estimate the elevation of the highest point in Florida, located in Walton County, if it is about 2.7% of the highest point in Arizona.

26. **BRAIN** The brain mass of a newborn baby is about 13% of the body mass of the newborn. If a newborn has a body mass of 2,900 grams, about how much is the mass of the brain?

27. **STOCKS** The value of a share of stock in an electronics company increased by $\frac{2}{3}$% during one week. If the value of a share of stock was $141 at the beginning of the week, estimate the increase in value of a share of stock at the end of the week.

Get ConnectED *For more practice, go to* <u>www.connected.mcgraw-hill.com</u>.

Multi-Part Lesson 1 PART C

Problem-Solving Practice

Percent and Estimation

---

1. **ORCHESTRA** The orchestra at Millard Middle School has 120 members. Of these, 17% are eighth-grade students. Estimate the number of eighth-grade students in the orchestra.

2. **RESTAURANTS** In one west coast city, 34% of the restaurants are on the river. The city has 178 restaurants. Estimate the number of restaurants that are on the river.

---

3. **FARMING** Rhonda planted green beans on 67% of her farm. Rhonda's farm has 598 acres of land. Estimate the number of acres of green beans on Rhonda's farm.

4. **HOTELS** At the Eastward Inn hotel, 47% of the rooms face the pool. The hotel has 92 rooms. Estimate the number of rooms that face the pool.

---

5. **TREES** The students in Leon's seventh grade science class determined that 42% of the trees at a local park are pine trees. If there are 632 trees in the park, about how many of them are pine trees?

6. **BOOKS** Jenna has read 0.7% of a book. If the book has 431 pages, estimate the number of pages Jenna has read.

---

7. **FITNESS** At the office where Mika works, 40% of the 18 employees exercise at least three times a week. Estimate the number of people who exercise at least three times a week.

8. **PETS** Of all seventh-grade students at Hart Middle School, 0.3% of the students own a pet iguana. If there are 610 seventh-grade students at Hart, about how many own pet iguanas?

Multi-Part Lesson 2

PART B

Homework Practice
The Percent Proportion

Find each number. Round to the nearest tenth if necessary.

1. What percent of 65 is 13?   2. $4 is what percent of $50?   3. What number is 35% of 22?

4. 14% of 81 is what number?   5. 13 is 26% of what number?   6. 55 is 40% of what number?

7. What percent of 45 is 72?   8. 1% of what number is 7?   9. 33 is 50% of what number?

10. What number is 3% of 100?   11. What percent of 200 is 0.5?

12. What number is 0.4% of 20?   13. What number is 6.1% of 60?

14. What percent of 34 is 34?   15. 10.4% of what number is 13?

16. **ALLOWANCE** Mallorie has $3 in her wallet. If this is 10% of her monthly allowance, what is her monthly allowance?

17. **WEDDING** Of the 125 guests invited to a wedding, 104 attended the wedding. What percent of the invited guests attended the wedding?

18. **CAMERA** The memory card on Melcher's digital camera can hold about 430 pictures. Melcher used 18% of the memory card while taking pictures at a family reunion. About how many pictures did Melcher take at the family reunion? Round to the nearest whole number.

19. **OCEANS** Use the table shown.

    a. The area of the Indian Ocean is what percent of the area of the Pacific Ocean? Round to the nearest whole percent.

    b. If the area of the Arctic Ocean is 16% of the area of the Atlantic Ocean, what is the area of the Arctic Ocean? Round to the nearest whole million.

Ocean	Area (square miles)
Pacific	64 million
Atlantic	32 million
Indian	25 million

Get ConnectED *For more practice, go to* www.connected.mcgraw-hill.com.

Problem-Solving Practice
The Percent Proportion

---

1. DRIVING Mudrik installed a device on his car that guaranteed to increase his gas mileage by 15%. He currently gets 22 miles per gallon. How much will the gas mileage increase after installing the device?

2. POPULATION The number of students at Marita's school decreased to 98% of last year's number. Currently, there are 1,170 students. How many students were there last year? Round to the nearest whole number.

---

3. VOTING Yolanda's club has 35 members. Its rules require that 60% of them must be present for any vote. At least how many members must be present to have a vote?

4. GARBAGE This month, Chun's office produced 690 pounds of garbage. Chun wants to reduce the weight of garbage produced to 85% of the weight produced this month. What is the target weight for the garbage produced next month?

---

5. SALARIES Yara just received a 6% raise in salary. Before the raise, she was making $52,000 per year. How much more will Yara earn next year?

6. SPORTS Sally's soccer team played 25 games and won 17 of them. What percent did the team win?

Homework Practice

The Percent Equation

Write an equation for each problem. Then solve. Round to the nearest tenth if necessary.

1. What number is 27% of 52?

2. Find 41% of 48.

3. What percent of 88 is 33?

4. 8 is what percent of 18?

5. What number is 33% of 360?

6. What percent of 62 is 58?

7. 55 is what percent of 100?

8. 22% of what number is 24.2?

9. 19 is 50% of what number?

10. 25 is 32% of what number?

11. 40% of what number is 28?

12. 30 is what percent of 60?

13. BASEBALL A baseball player was at bat 473 times during the regular season. If he made a hit 31.5% of the times he was at bat, how many hits did he make during the regular season? Round to the nearest whole number if necessary.

14. ANALYZE GRAPHS Use the graph shown. The total enrollment at Central High School is 798 students.

a. About what percent of the students at Central High are freshmen? Round to the nearest tenth if necessary.

b. About what percent of the students at Central High are seniors? Round to the nearest tenth if necessary.

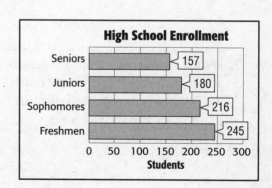

High School Enrollment

Seniors — 157
Juniors — 180
Sophomores — 216
Freshmen — 245

0 50 100 150 200 250 300
Students

Get ConnectED *For more practice, go to* www.connected.mcgraw-hill.com.

Problem-Solving Practice

The Percent Equation

1. **DINING** Jonas and Norma's restaurant bill comes to $23.40. They are planning to tip the waiter 15% of their bill. How much money should they leave for a tip?

2. **CHESS** The Briarwood Middle School chess club has 55 members. 22 of the members are in seventh grade. What percent of the members of the chess club are in seventh grade?

3. **TENNIS** In the city of Springfield, 75% of the parks have tennis courts. If 15 parks have tennis courts, how many parks does Springfield have altogether?

4. **COLLEGE** There are 225 students in eighth grade at Jefferson Middle School. A survey shows that 64% of them are planning to attend college. How many Jefferson eighth-grade students are planning to attend college?

5. **BASEBALL** In a recent season, the Chicago White Sox won 99 out of 162 games. What percent of games did the White Sox win? Round to the nearest tenth if necessary.

6. **HOUSING** In the Stoneridge apartment complex, 35% of the apartments have one bedroom. If there are 49 one-bedroom apartments, what is the total number of apartments at Stoneridge?

7. **SPACE** On Mars, an object weighs 38% as much as on Earth. How much would a person who weighs 165 pounds on Earth weigh on Mars?

8. **FOOTBALL** Javier had 2 passes intercepted out of 17 attempts in his last football game. What percent of Javier's passes were intercepted? Round to the nearest tenth if necessary.

Multi-Part
Lesson **2**

PART **D**

Homework Practice
Problem-Solving Investigation:
Determine Reasonable Answers

Mixed Problem Solving

Use the *determine reasonable answers* strategy for Exercises 1 and 2.

1. **HOMES** In a retirement village, 86% of the residents own their home. If the village has 540 homes, about how many homes are owned by the residents, about 250, 350, or 450?

2. **ANALYZE GRAPHS** The graph shows how the Forenzo family spent their money on their summer vacation. Is 25% a reasonable estimate of how much money they spent on dining? Justify your answer.

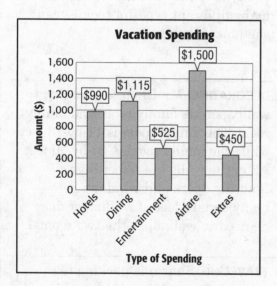

Use any strategy to solve Exercises 3–5.

3. **ANALYZE GRAPHS** The graph shows the percent of community attendance during a little league season. Is 90% a reasonable estimate for the percent of community attendance for September? Explain.

4. **TRAVEL** Cecil averages 31 miles per gallon when driving his car to visit friends 461 miles away. If he filled the 16-gallon gasoline tank before leaving and did not buy any gasoline along the way, about how many gallons of gasoline are left in the tank when he arrives? Justify your answer.

5. **FABRIC** Mrs. Tillman is making identical dresses for her three granddaughters. She needs $2\frac{1}{8}$ yards of fabric for each dress. If she purchased $8\frac{1}{2}$ yards of fabric, how much fabric will be leftover? Justify your answer.

Get ConnectED *For more practice, go to* www.connected.mcgraw-hill.com.

Problem-Solving Practice
Problem-Solving Investigation:
Determine Reasonable Answers

Solve using any method.

1. **GYM** The 6ᵗʰ graders are running the mile in physical education. Elian finishes the mile 2 minutes before Stacey who finished 1 minute 26 seconds behind Kareem. If Joanna completes the mile 1 minute and 42 seconds after Kareem, and her time is 8 minutes 34 seconds, what is Elian's time?

2. **POLITICS** A candidate receives 62% of the vote in an election and there are 1,603 votes recorded. How many votes did the candidate receive?

3. **POPULATION** The population of the United States is about 300,000,000. If Spanish is the primary language for 10.7% of the population, about how many people speak Spanish as their primary language?

4. **BAKING** Bea has prepared a basic cookie dough to which she will add ingredients to make several types of cookies. She has chocolate chips, raisins, and peanut butter chips. She also has peanuts, pecans, and walnuts. If she wants to put one ingredient from the first group with one type of nut into the dough, how many different types of cookies can she make?

5. **COINS** Zachary has four different coins that total 41 cents. What coins does he have?

6. **DECORATING** Mr. Chen is planning to wallpaper his family room and dining room. The dining room is 11-by-13 feet, while the family room is 20-by-10 feet. All of the walls are 8 feet high. Ignoring doors and windows, how many square feet of wallpaper does he need to wallpaper the two rooms?

7. **MOVIES** Charis is going to the movies with a friend. The price of admission is $5.50, a small popcorn is $2.39, and a small drink is $2.65. If Charis has a ten dollar bill, does she have enough money for admission, popcorn, and a drink? If not, how much more money would she need?

8. **TRAVELING** Shawn is packing his suitcase for vacation. If he has 2 pairs of shorts, and 5 shirts, how many different outfits can he make?

Homework Practice

Percent of Change

For Exercises 1–14, find each percent of change. Round to the nearest whole percent if necessary. State whether the percent of change is an *increase* **or** *decrease.*

1. 8 feet to 10 feet   **2.** 136 days to 85 days   **3.** $0.32 to $0.37

4. 62 trees to 31 trees   **5.** 51 meters to 68 meters   **6.** 16.5 grams to 24.8 grams

7. 0.55 minute to 0.1 minute   **8.** $180 to $210

9. 2.9 months to 4.9 months   **10.** 0.5 to 0.75

11. 0.1 to 0.2   **12.** 1.5 to 0.375

13. SURGERY Recent developments in surgical procedures change the average healing time for some operations from 8 weeks to 3 weeks.

14. ROADS The city added an extra lane in each direction to the 5-lane road.

15. GEOMETRY Refer to the rectangle shown. Suppose the width of 4 inches is decreased by 3 inches.
  a. Find the percent of change in the perimeter.

  b. Find the percent of change in the area.

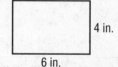

4 in.

6 in.

16. ANALYZE TABLES Refer to the table that shows the average monthly rainfall during the first six months of the year for Singapore.

  a. Between which two consecutive months is the percent of decrease the greatest? What is the percent change to the nearest whole percent?

  b. Between which two consecutive months is the percent of increase the least? What is the percent change to the nearest whole percent?

Month	Average Rainfall (inches/month)
January	9.4
February	6.5
March	6.8
April	6.6
May	6.7
June	6.4

Get ConnectED *For more practice, go to* <u>www.connected.mcgraw-hill.com</u>.

Multi-Part Lesson 3
PART B

Problem-Solving Practice

Percent of Change

1. **SHOES** A popular brand of running shoes costs a local store $68 for each pair. If the store sells the shoes for $119, what is the percent of increase in the price?

2. **CLUBS** Last year the backgammon club had 30 members. This year the club has 24 members. Find the percent of decrease in the number of members.

3. **READING** In the seventh grade, Rachel read 15 books. In the eighth grade, she read 18 books. Find the percent of increase in the number of books Rachel read.

4. **VOTES** Last year 762 students voted in the student council election at San Bruno Middle School. This year 721 students voted. What was the percent of change in the number of students that voted to the nearest tenth?

5. **HEIGHT** When Ricardo was 9 years old, he was 56 inches tall. Ricardo is now 12 years old and he is 62 inches tall. Find the percent of increase in Ricardo's height to the nearest tenth.

6. **PLANTS** Alicia planted 45 tulip bulbs last year. This year she plans to plant 65 bulbs. Find the percent of increase in the number of tulip bulbs to the nearest tenth.

7. **PICTURES** The 2008 yearbook at Middleton Middle School had 236 candid pictures of students. The 2007 yearbook had 214 candid pictures of students. What was the percent of change in the number of candid student pictures from 2007 to 2008 to the nearest tenth?

8. **POPULATION** In 2000, Florida's population was 15.9 million. In 2008, the population was estimated to be 18.8 million. Find the percent of increase to the nearest whole number.

Course 2 • Percent

Multi-Part Lesson 3

PART C

Homework Practice

Sales Tax and Tips

Find the total cost to the nearest cent.

1. $18.00 breakfast; 7% tax

2. $14 meal; 20% tip

3. $24 lunch; 15% tip

4. $8.50 shorts; 6.5% tax

5. $75 dinner; 18% tip

6. $74.95 jacket; 5% tax

7. $185 DVD player; 6% tax

8. $85 jeans; 7% tax

9. $20 haircut; 10% tip

10. $7.95 lunch; 15% tip

11. **MEAL** Enrique took his family out for dinner. He planned to leave a 15% gratuity on the bill. What is the total cost if the bill was $123.50?

12. **TRUCKS** What is the sales tax on a $17,500 truck if the tax rate is 6%?

13. **COMPUTER** Lionel is buying a computer that normally sells for $890. The state sales tax rate is 6%. What is the total cost of the computer including sales tax?

Get ConnectED *For more practice, go to* www.connected.mcgraw-hill.com.

Problem-Solving Practice

Sales Tax and Tips

1. SKATEBOARDS Inez wants to buy a skateboard but she does not know if she has enough money. The price of the skateboard is $80 and the sales tax is 7%. What will be the total cost of the skateboard?

2. HAIRCUT Josiah went to the local barber to get his hair cut. It cost $18 for the haircut. Josiah tipped the barber 15%. What was the total cost of the haircut including the tip?

3. MEAL Madeline took 3 friends out for dinner. The cost of the meals was $46.50. She left a 20% tip. What was the total cost including the tip?

4. COMPUTERS Andrea ordered a computer on the Internet. The computer cost $1,399 plus $6\frac{1}{2}\%$ sales tax. What was the total amount Andrea paid for her computer?

5. MAGAZINES Ivan bought these two magazines. If the sales tax was 6.75%, what was the total amount that he paid for the magazines?

$4.95 $4.95

6. CATERED DINNER The Striton family had a meal catered for a wedding rehearsal dinner. The cost of the dinner was $476. There was a 5% sales tax and they left a 15% tip. What was the total cost including the sales tax and the tip?

Multi-Part Lesson **3**
PART **D**

Homework Practice
Discount

Find the sale price to the nearest cent.

1. $239 television; 10% discount

2. $72 game; 20% discount

3. $18.95 football; 15% discount

4. $10.99 CD; 25% discount

5. $149 MP3 player; 40% discount

6. $213 ski jacket; 30% discount

7. $595 refrigerator; 20% discount

8. $64 video game; 25% discount

9. $119 croquet set; 50% discount

10. $14.99 clock; 10% discount

11. **RADIO** A radio is on sale for $50. If this price represents a 10% discount from the original price, what is the original price to the nearest nickel?

12. **LAUNDRY DETERGENT** A box of laundry detergent is on sale for $6.50. If this price represents a 40% discount from the original price, what is the original price to the nearest cent?

13. **BASKETBALL** Find the price of a $35 basketball that is on sale for 50% off the regular price.

Get ConnectED *For more practice, go to* www.connected.mcgraw-hill.com.

Multi-Part Lesson **3**

PART **D**

Problem-Solving Practice

Discount

1. **PRETZELS** The Spanish club sold hot pretzels as a fundraiser. The pretzels normally sold for $2.00, but near the end of the sale the price was reduced by 25%. What was the new price for a hot pretzel?

2. **CELL PHONES** Nathan is buying a cell phone for his business. The regular price of the cell phone is $179. If he buys the phone in the next 2 weeks, he will get a 20% discount. What will be the sale price if he buys the phone tomorrow?

3. **ALARM CLOCK** Dominic bought a new alarm clock that was on sale for $18.75. If this price represents a 30% discount from the original price, what is the original price to the nearest cent?

4. **FISHING ROD** Malachi bought a new fishing rod. The regular price of the fishing rod was $125.99. He bought it on sale with a 15% discount. Sales tax of 3% is applied to the discounted total. What was the sale price with tax of Malachi's fishing rod to the nearest cent?

5. **JEWELRY** A jewelry store is having a 50% off sale for all necklaces. During this sale, what is the cost of a necklace that regularly cost $49.98?

6. **COSMETICS** Jaylynn was buying new mascara. She bought it on sale for $5.56. If the price represents a 20% discount from the original price, what is the original price to the nearest cent?

Multi-Part
Lesson **3**
PART E

Homework Practice

Financial Literacy: Simple Interest

Find the simple interest earned to the nearest cent for each principal, interest rate, and time.

1. $750, 7%, 3 years

2. $1,200, 3.5%, 2 years

3. $450, 5%, 4 months

4. $1,000, 2%, 9 months

5. $530, 6%, 1 year

6. $600, 8%, 1 month

Find the simple interest paid to the nearest cent for each loan, interest rate, and time.

7. $668, 5%, 2 years

8. $720, 4.25%, 3 months

9. $2,500, 6.9%, 6 months

10. $500, 12%, 18 months

11. $300, 9%, 3 years

12. $2,000, 20%, 1 year

13. ELECTRONICS Rita charged $126 for a DVD player at an interest rate of 15.9%. How much will Rita have to pay after 2 months if she makes no payments?

14. VACATION A family invests $1,050 for a vacation at an interest rate of 11.9%. After 6 months, what is the total amount that the family has saved for the vacation?

15. INVESTMENTS Serena has $2,500 to invest in a CD (certificate of deposit).

   a. If Serena invests the $2,500 in the CD that yields 4% interest, what will the CD be worth after 2 years?

   b. Serena would like to have $3,000 altogether. If the interest rate is 5%, in how many years will she have $3,000?

   c. Suppose Serena invests the $2,500 for 3 years and earns $255. What was the rate of interest?

Get ConnectED *For more practice, go to www.connected.mcgraw-hill.com.*

Problem Solving Practice

Financial Literacy: Simple Interest

1. SAVINGS ACCOUNT How much interest will Hannah earn in 4 years if she deposits $630 in a savings account at 6.5% simple interest?

2. INVESTMENTS Terry invested $2,200 in the stock market for 2 years. If the investment earned 12%, how much money did Terry earn in 2 years?

3. RETIREMENT Mr. Pham has $410,000 in a retirement account that earns 3.85% simple interest each year. Find the amount earned each year by this investment.

4. COLLEGE FUND When Melissa was born, her parents put $8,000 into a college fund account that earned 9% simple interest. Find the total amount in the account after 18 years.

5. INHERITANCE Raj inherited $900,000 from his father. After paying $350,000 for a house, he invested the remaining money in a savings account at 4.25% simple interest. How much money is in the account if Raj makes no deposits or withdrawals for two years?

6. SAVINGS Mona opened a savings account with a $500 deposit and a simple interest rate of 5.6%. If there were no deposits or withdrawals, how much money is in the account after $8\frac{1}{2}$ years?

7. SAVINGS ACCOUNT Malik deposited $1,050 in a savings account, and it earned $241.50 in simple interest after four years. Find the interest rate on Malik's savings account.

8. INHERITANCE Kelli Rae's inheritance from her great-grandmother was $220,000 after taxes. If Kelli Rae invests this money in a savings account that earns $18,260 in simple interest every year, what is the interest rate on her account?

Homework Practice

Equations and Functions

Complete each function table. Then identify the domain and range.

1. $y = 5x$

x	5x	y
1		
2		
3		
4		

2. $y = 8x$

x	8x	y
1		
2		
3		
4		

3. $y = 7x$

x	7x	y
3		
4		
5		
6		

4. $y = -4x$

x	-4x	y
2		
3		
4		
5		

5. $y = \frac{1}{2}x$

x	$\frac{1}{2}x$	y
2		
4		
6		
8		

6. $y = 0.75x$

x	0.75x	y
0		
2		
4		
6		

7. PRODUCTION A car manufacturer makes 15,000 hybrid cars a month. Using the function table, find the number of hybrid cars produced after 3, 6, 9, and 12 months.

m	15,000m	P
3		
6		
9		
12		

8. SUNSPOTS The changing activity of sunspots, which are cooler and darker areas of the sun, occur in 11-year cycles. Use the function $y = 11c$ to find the numbers of years necessary to complete 1, 2, 3, and 4 sunspot cycles.

Get ConnectED *For more practice, go to* www.connected.mcgraw-hill.com.

Problem-Solving Practice

Equations and Functions

1. **TECHNOLOGY** The fee for your pager service is $22 per month. Make a function table that shows your total charge for 1, 2, 3, and 4 months of service.

2. **MEASUREMENT** Josef takes 2 steps for every one step that Kim takes. Write an equation in two variables showing the relationship between Josef's steps and Kim's steps. If Kim takes 15 steps, how many steps will Josef have to take to cover the same distance?

3. **TRAINS** Between Hiroshima and Kokura, Japan, the bullet train averages a speed of 164 miles per hour, which is the fastest scheduled train service in the world. Make a function table that shows the distance traveled at that speed in 1, 2, 3, and 4 hours.

4. **BUSINESS** Grant earns $5 for each magazine that he sells. Write an equation in two variables showing the relationship between the number of magazines sold and the amount of money made. If Grant sells 12 magazines, how much money will he make?

5. **GEOMETRY** The formula for the volume of a rectangular prism whose base has an area of 8 square units is $V = 8h$, where V is the volume and h is the height. Make a function table that shows the volume of a rectangular prism with a height of 3, 4, 5, and 6 units.

6. **ANIMALS** A dragonfly can fly at 36 miles per hour. Write an equation in two variables describing the relationship between the length of the dragonfly's flight and the distance traveled. If a dragonfly flies for 3 hours, how far can he travel?

Multi-Part Lesson 1

PART C

Homework Practice

Functions and Graphs

Graph each equation.

1. $y = x - 2$

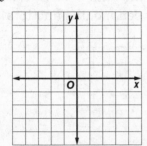

2. $y = -x$

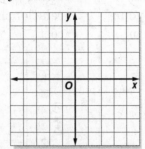

3. $y = 2x - 1$

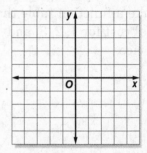

4. $y = 1.5x$

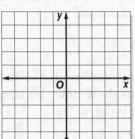

5. $y = x - 0.5$

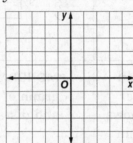

6. $y = 0.5x + 2$

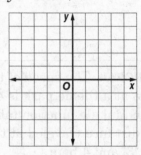

Graph the function represented by each table.

7.

x	y
0	3.5
1	2.5
2	1.5
3	0.5

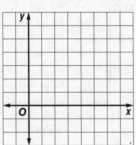

8.

x	y
1	6
0	4.5
−1	3
−2	1.5

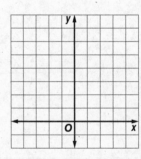

9. PRESSURE Ocean pressure increases about one atmosphere for every 10 meters of water depth. This can be represented by the function $p = 0.1d$ where p represents the pressure in atmospheres at a depth d. Represent this function by a graph.

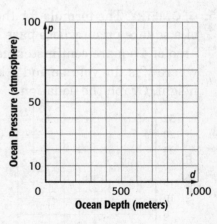

Get ConnectED *For more practice, go to* www.connected.mcgraw-hill.com.

Multi-Part
Lesson **1**

PART **C**

Problem-Solving Practice

Functions and Graphs

1. **TEXT MESSAGES** The cost for your text messaging package is $9 per month. Make a function table that shows your total cost for 1, 2, 3, and 4 months of service.

2. **TEXT MESSAGES** Refer to Exercise 1. Make a graph of the data to show the relationship between months and total cost.

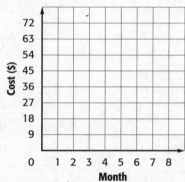

3. **MONORAIL** The speed of the monorail train at a theme park is 45 miles per hour. Make a function table that shows the distance traveled at that speed in 1, 2, 3, and 4 hours.

4. **MONORAIL** Refer to Exercise 3. Make a graph of the data to show how the time and distance are related.

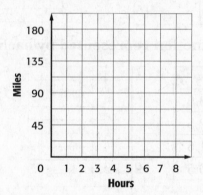

5. **GEOMETRY** The formula for the volume of a square prism whose base has an area of 4 square centimeters is $V = 4h$, where V is the volume and h is the height. Graph the equation.

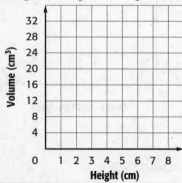

6. **BIRDS** The spine-tailed swift can fly at a speed of 106 miles per hour. Write an equation using x to represent hours and y to represent distance. Then graph the equation.

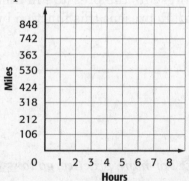

Multi-Part Lesson 2
PART B

Homework Practice

Constant Rate of Change

Find the constant rate of change for each table.

1.

Number of Pounds of Ham	Cost ($)
0	0
3	12
6	24
9	36

2.

Number of Hours Worked	Money Earned ($)
4	80
6	120
8	160
10	200

3.

Days	Plant Height (in.)
7	4
14	11
21	18
28	25

4.

Months	Money Spent on Cable TV
2	82
4	164
6	246
8	328

Find the constant rate of change for each graph.

5.

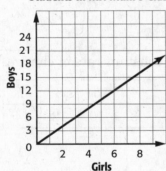

Students in Mr. Muni's Class

6.

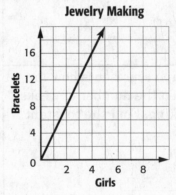

Jewelry Making

7. SEAGULLS At 1 P.M., there were 16 seagulls on the beach. At 3 P.M., there were 40 seagulls. What is the constant rate of change?

Get ConnectED *For more practice, go to* www.connected.mcgraw-hill.com.

Multi-Part Lesson 2
PART B

Problem-Solving Practice

Constant Rate of Change

1. **WATER** At 2 P.M., the level of the water in the pool was 10 feet. At 6 P.M., the level of water was 2 feet. Find the constant rate of change of the water.

2. **MONEY** JoAnne is depositing money into a bank account. After 3 months there is $150 in the account. After 6 months, there is $300 in the account. Find the constant rate of change of the account.

3. **TEMPERATURE** The temperature at noon was 88°F. By 4 P.M., the temperature was 72°F. Find the constant rate of change of the temperature.

4. **GROWTH** Jaz was 43 inches tall. Eighteen months later, she was 52 inches tall. Find the constant rate of change for Jaz's height.

5. **BIKING** The graph represents how far Toby biked given the number of weeks he has been biking. Find the constant rate of change.

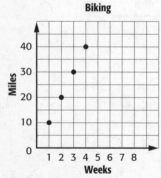

Biking

6. **HAIR** Find the constant rate of change.

Months	4	5	6	7
Length (in.)	8	10	12	14

Multi-Part Lesson 2

PART C

Homework Practice

Slope

For Exercises 1 and 2, graph the data. Then find the slope. Explain what the slope represents.

1. **ENVELOPES** The table shows the number of envelopes stuffed for various times.

Time (min)	5	10	15	20
Envelopes Stuffed	30	60	90	120

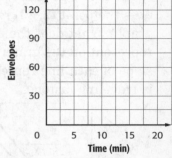

2. **MEASUREMENT** There are 3 feet for every yard.

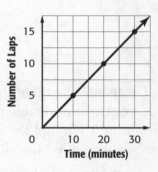

3. Use the graph that shows the number of laps completed over time. Find the slope of the line.

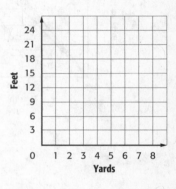

4. Which line is the steepest? Explain using the slopes of lines ℓ, m, and n.

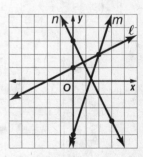

Get ConnectED *For more practice, go to www.connected.mcgraw-hill.com.*

Problem-Solving Practice

Slope

1. GO-KARTS The graph shows the average speed of two go-karts in a race. What does the point (2, 20) represent on the graph? What does the point (1, 12) represent on the graph? What does the slope of each line represent? Which car is traveling faster?

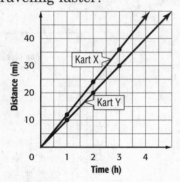

2. SKATEBOARDING The line represents the length and height of a skateboard ramp. Find the slope of the ramp.

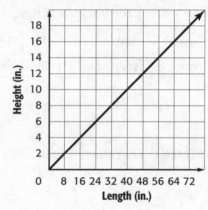

3. RAISINS The table shows the number of packages of raisins per box. Graph the data. Then find the slope of the line. Explain what the slope represents.

Packages	20	40	60	80
Boxes	1	2	3	4

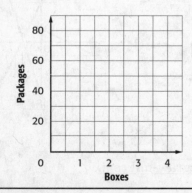

4. RESORT The Snells can spend 4 nights at a resort for $500 or 6 nights at the same resort for $750. Graph the data. Then find the slope. Explain what the slope represents.

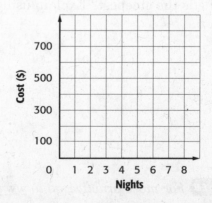

Course 2 • Linear Functions

Multi-Part Lesson 3
PART A

Homework Practice

Problem-Solving Investigation: Use a Graph

Mixed Problem Solving

Use a graph to solve Exercise 1.

1. **DRIVING** Ms. Bonilla recorded the amount of time it took her to drive to work each morning.

Day	Departure Time (A.M.)	Travel Time (min)
1st Week Monday	7:21	17
1st Week Tuesday	7:38	26
1st Week Wednesday	7:32	22
1st Week Thursday	7:20	15
1st Week Friday	7:35	22
2nd Week Monday	7:26	20
2nd Week Tuesday	7:25	18
2nd Week Wednesday	7:38	24
2nd Week Thursday	7:34	21
2nd Week Friday	7:23	17

a. Make a graph of the data.

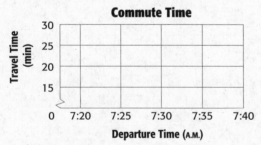

b. Predict the travel time if Ms. Bonilla leaves at 7:42.

Use any strategy to solve Exercises 2 and 3.

2. **FLORIST** Ms. Parker charges $29.95 for a bouquet of one dozen roses. Last year, she paid her supplier $4.50 per dozen roses. This year, she paid $3.25 more per dozen. How much less profit did she make this year on 20 dozen bouquets?

3. **TOURS** Use the graph showing the amount of money a tour bus company received from passengers and their cost of operations.

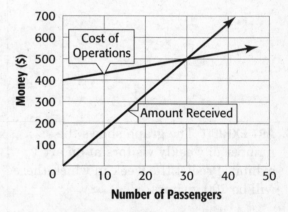

a. Predict the amount the tour bus company will receive if 50 passengers ride.

b. Predict the cost of operations if 50 passengers ride.

c. Use the values in parts **a** and **b** to predict the tour bus company's profit if 50 passengers ride.

Get ConnectED *For more practice, go to* www.connected.mcgraw-hill.com.

Problem-Solving Practice

Problem-Solving Investigation: Use a Graph

1. SALES The graph shows the monthly sales of George's Comic Book Shop. Predict the sales in July.

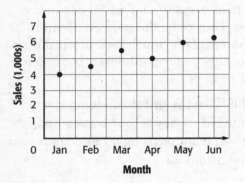

2. EXERCISING Ogden runs the mile race at every track meet. The table shows his times for each meet. Draw a graph. Predict Ogden's time in Meet 6.

Meet	1	2	3	4	5
Time	8:50	8:44	8:35	8:30	8:28

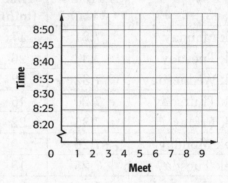

3. ART EXHIBIT The graph shows the number of weekly visitors at an art exhibit. Predict the week in which there will be 700 visitors.

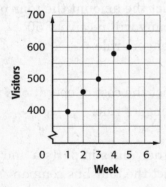

4. ART EXHIBIT Use the graph in Exercise 3 to estimate the constant rate of change per week of visitors at the art exhibit.

Multi-Part
Lesson **3**

PART **C**

Homework Practice

Direct Variation

1. **HOME THEATER** The number of home theaters a company sells varies directly as the money spent on advertising. How many home theaters does the company sell for each $500 spent on advertising?

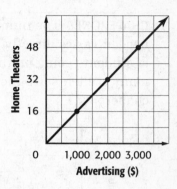

2. **DUNE BUGGY** Beach Travel rents dune buggies for $50 for 4 hours or $75 for 6 hours. What is the hourly rate?

3. **FERTILIZER** Leroy uses 20 pounds of fertilizer to cover 4,000 square feet of his lawn and 50 pounds to cover 10,000 square feet. How much does he need to cover his entire yard which has an area of 26,400 square feet?

Determine whether each linear function is a direct variation. If so, state the constant of variation.

4.

Gallons, x	6	8	10	12
Miles, y	180	240	300	360

5.

Time (min), x	10	11	12	13
Temperature, y	82	83	84	85

6.

Number of Payments, x	6	11	16	21
Amount Paid, y	$1,500	$2,750	$4,000	$5,250

If y varies directly with x, write an equation for the direct variation. Then find each value.

7. If $y = -4$ when $x = 10$, find y when $x = 5$.

8. If $y = 12$ when $x = -15$, find y when $x = 2$.

9. Find x when $y = 18$, if $y = 9$ when $x = 8$.

[Get ConnectED] *For more practice, go to* www.connected.mcgraw-hill.com.

Problem-Solving Practice

Direct Variation

1. CHAPERONES The number of chaperones needed varies directly with the number of students going on the trip. How many students are there for each chaperone?

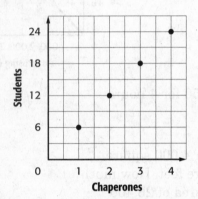

2. CANDLES The number of votive candles varies directly as the price. What is the ratio of candles to dollars?

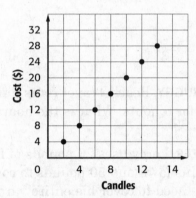

3. BAKING The number of cups of flour varies directly as the number of rolls made. Four cups of flour are needed to make 24 rolls. How much flour is needed for 36 rolls?

4. DISTANCE The number of feet between 2 objects varies directly as the number of miles. If the distance between 2 objects is $2\frac{1}{2}$ miles or 13,200 feet, how many feet are equivalent to a distance of 7 miles?

5. CHEWING GUM The table shows the number of sticks of chewing gum per pack. Find the number of sticks per pack.

Sticks	10	20	30	40
Packs	1	2	3	4

6. WEDDING FAVORS Lucius is making favors for his sister's wedding. If supplies for 25 favors cost $62.50, how much do supplies for 60 favors cost?

Multi-Part Lesson 3

PART E

Homework Practice

Inverse Variation

1. OCEAN The table shows the relationship between the temperature of the ocean and its depth. Graph the data in the table and determine if the relationship is an inverse variation.

Temperature (°C)	4	8	40	50
Depth (m)	1,000	500	100	80

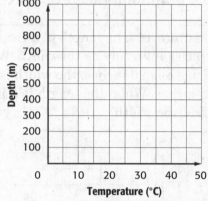

2. CONDUIT The table shows the relationship between the velocity of a fluid flowing in a conduit and the area of a cross section of the conduit. Graph the data in the table and determine if the relationship is an inverse variation.

Velocity (ft/s)	25	50	100
Area (in²)	2	1	$\frac{1}{2}$

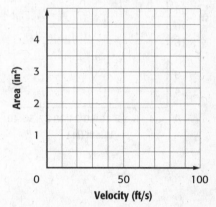

3. DISTANCE The table shows the relationship between the number of paper cups ordered and the cost per 100 cups. Graph the data in the table and determine if the relationship is an inverse variation.

Cups Ordered	10,000	7,500	6,000
Cost Per 100 Cups ($)	3	4	5

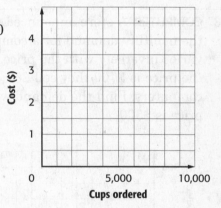

4. WAGES Mr. Anschutz's annual salary is shown in the table. Graph the data in the table and determine if the relationship is an inverse variation.

Year	2002	2005	2008	2011
Salary ($)	34,000	38,000	42,000	46,000

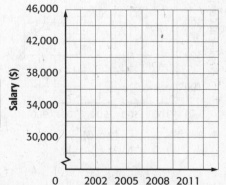

Get ConnectED *For more practice, go to* www.connected.mcgraw-hill.com.

Problem-Solving Practice

Inverse Variation

Draw a graph for each exercise.

1. **RECTANGLE** The length of a rectangle varies inversely as the width. When the length is 20 inches, the width is 2 inches. When the length is 5 inches, find the width.

2. **DRIVING** Ian can drive to the aquarium in 3 hours at 60 miles per hour. If speed varies inversely with times, how long will it take him at 36 miles per hour?

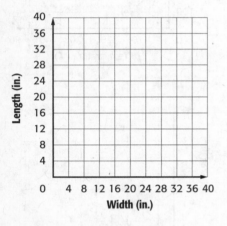

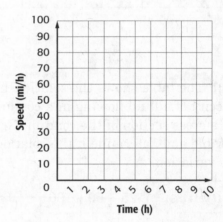

3. **COMPUTER** A store announced that the monthly demand for a computer varies inversely with the price. When the price is $700 they sell 250 computers. Find the demand when the price is $500.

4. **SEESAW** Ozzie and his friend are playing on a seesaw. Ozzie weighs 100 pounds and is sitting 10 feet from the balance point. If weight varies inversely with the distance from the balance point, where should Angie sit to balance him if she weighs 80 pounds?

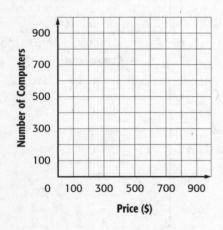

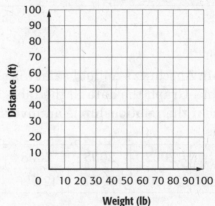

Multi-Part Lesson 1

PART A

Homework Practice

Probability and Simple Events

A set of cards is numbered 1, 2, 3, ..., 24. Suppose you pick a card at random without looking. Find the probability of each event. Write as a fraction in simplest form.

1. $P(5)$

2. P(multiple of 4)

3. P(6 or 17)

4. P(*not* equal to 15)

5. P(*not* a factor of 6)

6. P(odd number)

COMMUNITY SERVICE The table shows the students involved in community service. Suppose one student out of the 15 girls and 25 boys is randomly selected to represent the school at a state-wide awards ceremony. Find the probability of each event. Write as a fraction in simplest form.

Community Service	
6th graders	20
7th graders	8
8th graders	12

7. P(boy)

8. P(*not* 6th grader)

9. P(girl)

10. P(8th grader)

11. P(boy or girl)

12. P(6th or 7th grader)

13. P(7th grader)

14. P(*not* a 9th grader)

MENU A delicatessen serves different menu items, of which 2 are soups, 6 are sandwiches, and 4 are salads. How likely is it for each event to happen if you choose one item at random from the menu? Explain your reasoning.

15. P(sandwich)

16. P(*not* a soup)

17. P(salad)

18. NUMBER CUBE What is the probability of rolling an even number or a prime number on a number cube? Write as a fraction in simplest form.

19. CLOSING TIME At a convenience store there is a 25% chance a customer enters the store within one minute of closing time. Describe the complementary event and find its probability.

Get ConnectED *For more practice, go to* www.connected.mcgraw-hill.com.

Problem-Solving Practice

Probability and Simple Events

COINS Kyra opened her piggy bank and counted the number of each coin. The table at the right shows the results. For Exercises 1–3, assume that the coins are put in a bag and one is chosen at random.

Coin	Number
quarters	15
dimes	21
nickels	22
pennies	32

---

1. What is the probability that a quarter is chosen?

2. What is the probability that a nickel or a dime is chosen?

---

3. What is the probability that the chosen coin is worth more than 5 cents?

4. **NUMBER CUBES** Juan has a number cube, with faces numbered 1, 2, ..., 6. What is the probability that he rolls a 1 or a 5?

---

5. **SKATEBOARDS** Carlotta bought a new skateboard for which the probability of having a defective wheel is 0.015. What is the probability of *not* having a defective wheel?

6. **CALCULATORS** Kenny's teacher has 6 scientific calculators and 8 graphing calculators. If the teacher chooses one calculator at random, what is the probability that it will be a graphing calculator?

---

7. **VEHICLES** The rental car company had 14 sedans and 8 minivans available to rent. If the next customer picks a vehicle at random, what is the probability that a minivan is chosen?

8. **MUSIC** Felisa has 16 pop CDs, 6 classical, and 2 rock. Felisa chooses a CD at random. What is the probability she does *not* choose a classical CD?

Multi-Part Lesson 1

PART B

Homework Practice

Sample Spaces

For Exercises 1 and 2, find the sample space using a tree diagram.

1. choosing blue, green, or yellow wall paint with white, beige, or gray curtains

2. choosing a lunch consisting of a soup, salad, and sandwich from the menu shown in the table

Soup	Salad	Sandwich
Tortellini	Caesar	Roast Beef
Lentil	Macaroni	Ham
		Turkey

3. **GAME** Kimiko and Miko are playing a game in which each girl rolls a number cube. If the sum of the numbers is a prime number, then Miko wins. Otherwise Kimiko wins. The sample space is shown. Find the probability that Kimiko wins.

Sum = 2	Sum = 3	Sum = 4	Sum = 5	Sum = 6	Sum = 7	Sum = 8	Sum = 9	Sum = 10	Sum = 11	Sum = 12
$1 + 1 = 2$	$2 + 1 = 3$ $1 + 2 = 3$	$1 + 3 = 4$ $2 + 2 = 4$ $3 + 1 = 4$	$1 + 4 = 5$ $2 + 3 = 5$ $3 + 2 = 5$ $4 + 1 = 5$	$1 + 5 = 6$ $2 + 4 = 6$ $3 + 3 = 6$ $4 + 2 = 6$ $5 + 1 = 6$	$1 + 6 = 7$ $2 + 5 = 7$ $3 + 4 = 7$ $4 + 3 = 7$ $5 + 2 = 7$ $6 + 1 = 7$	$2 + 6 = 8$ $3 + 5 = 8$ $4 + 4 = 8$ $5 + 3 = 8$ $6 + 2 = 8$	$3 + 6 = 9$ $4 + 5 = 9$ $5 + 4 = 9$ $6 + 3 = 9$	$4 + 6 = 10$ $5 + 5 = 10$ $6 + 4 = 10$	$5 + 6 = 11$ $6 + 5 = 11$	$6 + 6 = 12$

For more practice, go to www.connected.mcgraw-hill.com.

Problem-Solving Practice

Sample Spaces

---

1. **GASOLINE** Craig stops at a gas station to fill his gas tank. He must choose between full-service or self-service and between regular, mid-grade, and premium gasoline. Draw a tree diagram showing the possible combinations of service and gasoline type. How many possible combinations are there?

2. **COINS** Lorelei tosses a coin 4 times. Draw a tree diagram showing the possible outcomes. What is the probability of getting at least 2 tails?

---

3. **COINS** In Exercise 2, what is the probability of getting 2 heads followed by 2 tails?

4. **EQUIPMENT** The computer accessory that Grace is considering selling at her store comes in white, beige, gray, or black and has an optical mouse, mechanical mouse, or trackball. How many combinations of color and model must she stock in order to have at least one of every possible combination?

Multi-Part Lesson 1

PART C

Homework Practice

Count Outcomes

Use the Fundamental Counting Principle to find the total number of outcomes in each situation.

1. choosing from 8 car models, 5 exterior paint colors, and 2 interior colors

2. selecting a year in the last decade and a month of the year

3. picking from 3 theme parks and 1-day, 2-day, 3-day, and 5-day passes

4. choosing a meat and cheese sandwich from the list shown in the table

Cheese	Meat
Provolone	Salami
Swiss	Turkey
American	Tuna
Cheddar	Ham

5. tossing a coin and rolling 3 number cubes

6. selecting coffee in regular or decaf, with or without cream, and with or without sweeteners

7 **COINS** Find the number of possible outcomes if 2 quarters, 4 dimes, and 1 nickel are tossed.

8. **SOCIAL SECURITY** Find the number of possible 9-digit social security numbers if the digits may be repeated.

9. **AIRPORTS** Jolon will be staying with his grandparents for a week. There are four flights that leave the airport near Jolon's home that connect to an airport that has two different flights to his grandparents' hometown. Find the number of possible routes. Then find the probability of taking the route with the earliest if the route is selected at random.

10. **ANALYZE TABLES** The table shows the kinds of homes offered by a residential builder. If the builder offers a discount on one home at random, find the probability it will be a 4-bedroom home with an open porch. Explain your reasoning.

Number of Bedrooms	Style of Kitchen	Type of Porch
5-bedroom	Mediterranean	Open
4-bedroom	Contemporary	Screen
3-bedroom	Southwestern	

Get ConnectED *For more practice, go to* www.connected.mcgraw-hill.com.

Problem-Solving Practice

Count Outcomes

1. SURFBOARD Tradd owns 3 surfboards and 2 wet suits. If he takes one surfboard and one wet suit to the beach, how many different combinations can he choose?	**2. SHOPPING** Trey is trying to decide which bag of dog food to buy. The brand he wants comes in 4 flavors and 3 sizes. How many choices are there?
3. PASSWORDS To set a password, you must select 4 numbers from 0 to 9. How many possible passwords can be chosen if each number may be used more than once?	**4. RESTAURANTS** Margaret's favorite restaurant has 3 specials every day. There are 2 choices of vegetable and 3 choices of dessert. How many different meals could Margaret have if she chooses one special, one vegetable, and one dessert?
5. ROUTES When Sunil goes to the building where he works, he can go through 4 different doors into the lobby. Then he can go to the seventh floor by taking 2 different elevators or 2 different stairways. How many different ways can Sunil get from outside the building to the seventh floor?	**6. STEREOS** Jailin went to her local stereo store. Given her budget and the available selection, she can choose between 2 CD players, 5 amplifiers, and 3 pairs of speakers. How many different ways can Jailin choose one CD player, one amplifier, and one pair of speakers?
7. DESSERT For dessert you can choose apple, cherry, blueberry, or peach pie to eat, and milk or juice to drink. How many different combinations of one pie and one beverage are possible?	**8. TESTS** Giorgio is taking a true or false quiz. There are six questions on the quiz. How many ways can the quiz be answered?

Homework Practice

Permutations

Solve each problem.

1. NUMBERS How many different 2-digit numbers can be formed from the digits 4, 6, and 8? Assume no number can be used more than once.

2. LETTERS How many permutations are possible of the letters in the word *numbers*?

3. PASSENGERS There are 5 passengers in a car. In how many ways can the passengers sit in the 5 passenger seats of the car?

4. PAINTINGS Mr. Bernstein owns 14 paintings, but has only enough wall space in his home to display three of them at any one time. How many ways can Mr. Bernstein display three paintings in his home?

5. DOG SHOW Mateo is one of the six dog owners in the terrier category. If the owners are selected in a random order to show their dogs, how many ways can the owners show their dogs?

6. TIME Michel, Jonathan, and two of their friends each ride their bikes to school. If they have an equally-likely chance of arriving first, what is the probability that Jonathan will arrive first and Michel will arrive second?

7. BIRTHDAY Glen received 6 birthday cards. If he is equally likely to read the cards in any order, what is the probability he reads the card from his parents and the card from his sister before the other cards?

CODES For Exercises 8–10, use the following information. A bank gives each new customer a 4-digit code number which allows the new customer to create their own password. The code number is assigned randomly from the digits 1, 3, 5, and 7, and no digit is repeated.

8. What is the probability that the code number for a new customer will begin with a 7?

9. What is the probability that the code number will *not* contain a 5?

10. What is the probability that the code number will start with 371?

Get ConnectED *For more practice, go to* www.connected.mcgraw-hill.com.

Problem-Solving Practice
Permutations

---

1. AREA CODES How many different 3-digit area codes can be created if no digit can be repeated? Assume all area codes are possible.

2. CARDS Jason is dealt five playing cards. In how many different orders could Jason have been dealt the same hand?

---

3. PASSWORDS How many different 3-letter passwords are possible if no letter may be repeated?

4. RACING All 22 students in Amy's class are going to run the 100-meter dash. In how many ways can the students finish in first, second, and third place?

---

5. LETTERS How many ways can you arrange the letters in the word *history*?

6. SCHOOL In how many ways can a president, vice-president, and secretary be chosen from eight students?

---

7. SERIAL NUMBERS How many different 6-digit serial numbers are available if no digit can be repeated?

8. WINNERS There are 156 ways for 2 cars to win first and second place in a race. How many cars are in the race?

Homework Practice
Independent and Dependent Events

The two spinners at the right are spun. Find each probability.

1. P(4 and C)

2. P(1 and A)

3. P(even and C)

4. P(odd and A)

5. P(greater than 3 and B)

6. P(less than 5 and B)

GAMES There are 10 yellow, 6 green, 9 orange, and 5 red cards in a stack of cards turned facedown. Once a card is selected, it is *not* replaced. Find each probability.

7. P(two yellow cards)

8. P(two green cards)

9. P(a yellow card and then a green card)

10. P(a red card and then an orange card)

11. P(two cards that are *not* orange)

12. P(two cards that are neither red nor green)

13. **OFFICE SUPPLIES** A store sells a box of highlighters that contains 4 yellow, 3 blue, 2 pink, and 1 green highlighter. What is the probability of randomly picking first 1 blue and then 1 pink highlighter from the box?

14. **BASKETBALL** Angelina makes 70% of her free throws. If this pattern continues, what is the probability that she will make her next two free throws?

15. **CAR RENTALS** Use the following information and the information in the table.

    At a car rental office, 63% of the customers are men and 37% are women. Assume the pattern continues.

Car Requests	
Compact	25%
Full-size	37%
Convertible	10%
SUV	16%
Luxury	12%

   a. What is the probability that the next customer will be a woman who requests a convertible?

   b. What is the probability that the next customer will be a man who requests either a compact car or luxury car?

Get ConnectED *For more practice, go to* www.connected.mcgraw-hill.com.

Problem-Solving Practice
Independent and Dependent Events

1. **CHECKERS** In a game of checkers, there are 12 red game pieces and 12 black game pieces. Julio is setting up the board to begin playing. What is the probability that the first two checkers he pulls from the box at random will be two red checkers?

2. **CHECKERS** What is the probability that the first piece is red and the second piece is black? Explain how you found your answer.

CHESS For Exercises 3–5, use the following information.

Inger keeps her white and black chess pieces in separate bags. For each color, there are 8 pawns, 2 rooks, 2 bishops, 2 knights, 1 queen, and 1 king.

3. Are the events of drawing a knight from the bag of white pieces and drawing a pawn from the bag of black pieces *dependent* or *independent* events? Explain. Find the probability of this compound event.

4. Are the events of drawing a bishop from the bag of white pieces and then drawing the queen from the same bag *dependent* or *independent* events? Explain. Find the probability of this compound event.

5. Find the probability of drawing a pawn, a knight, and another pawn from the bag of white pieces.

6. **SOCCER** During a soccer season, Mario made approximately 2 goal points for every 5 of his shots on goal. What is the probability that Mario would make 2 goal points on two shots in a row during the season?

Course 2 • Probability and Predictions

Homework Practice

Theoretical and Experimental Probability

1. A number cube is rolled 24 times and lands on 2 four times and on 6 three times.

   a. Find the experimental probability of landing on a 2.

   b. Find the experimental probability of *not* landing on a 6.

   c. Compare the experimental probability you found in part a to its theoretical probability.

   d. Compare the experimental probability you found in part b to its theoretical probability.

2. **ENTERTAINMENT** Use the results of the survey in the table shown.

   a. What is the probability that someone in the survey considered reading books or surfing the Internet as the best entertainment value? Write the probability as a fraction.

   b. Out of 500 people surveyed, how many would you expect considered reading books or surfing the Internet as the best entertainment value?

   c. Out of 300 people surveyed, is it reasonable to expect that 30 considered watching television as the best entertainment value? Why or why not?

Best Entertainment Value	
Type of Entertainment	**Percent**
Playing Interactive Games	48
Reading Books	22
Renting Movies	10
Going to Movie Theaters	10
Surfing the Internet	9
Watching Television	1

3. A spinner marked with four sections blue, green, yellow, and red was spun 100 times. The results are shown in the table.

   a. Find the experimental probability of landing on green.

   b. Find the experimental probability of landing on red.

   c. If the spinner is spun 50 more times, how many of these times would you expect the pointer to land on blue?

Section	Frequency
Blue	14
Green	10
Yellow	8
Red	68

GetConnectED For more practice, go to www.connected.mcgraw-hill.com.

Problem Solving Practice

Theoretical and Experimental Probability

HOBBIES For Exercises 1–4, use the graph of a survey of 24 seventh-grade students asked to name their favorite hobby.

What is your favorite hobby?

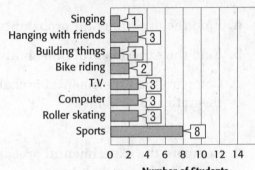

1. What is the probability that a student's favorite hobby is roller skating?

2. Suppose 200 seventh-grade students were surveyed. How many can be expected to say that roller skating is their favorite hobby?

3. Suppose 60 seventh-grade students were surveyed. How many can be expected to say that bike riding is their favorite hobby?

4. Suppose 150 seventh-grade students were surveyed. How many can be expected to say that playing sports is their favorite hobby?

WINTER ACTIVITIES For Exercises 5 and 6, use the graph of a survey with 104 responses in which respondents were asked about their favorite winter activities.

What is your favorite winter activity?

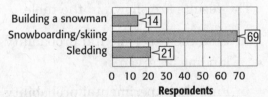

5. What is the probability that someone's favorite winter activity is building a snowman? Write the probability as a fraction.

6. If 500 people had responded, how many would have been expected to list sledding as their favorite winter activity? Round to the nearest whole person.

Multi-Part Lesson **3**

PART **C**

Homework Practice

Problem-Solving Investigation: Act it Out

Mixed Problem Solving

For Exercises 1 and 2, use the *act it out* strategy.

1. **POP QUIZ** Use the information in the table to determine whether tossing a nickel and a dime is a good way to answer a 5-question multiple-choice quiz if each question has answer choices A, B, C, and D. Justify your answer.

Nickel	Dime	Answer Choice
H	H	A
H	T	B
T	H	C
T	T	D

2. **BOWLING** Louis, Lucas, Garcelle, and Sheryl go bowling every week. When ordered from highest to lowest, how many ways can their scores be arranged if Lucas is never first and Garcelle always beats Louis?

Use any strategy to solve Exercises 3–7.

3. **BOOKS** What is the probability of five books being placed in alphabetical order by their titles if randomly put on a book shelf?

4. **NUMBER THEORY** The sum of a 2-digit number and the 2-digit number when the digits are reversed is 77. If the difference of the same two numbers is 45, what are the two 2-digit numbers?

5. **BASEBALL** In one game, Rafael was up to bat 3 times and made 2 hits. In another game, he was up to bat 5 times with no hits. What percent of the times at bat did Rafael make a hit?

6. **RESTAURANT** A restaurant offers the possibility of 168 three-course dinners. Each dinner has an appetizer, an entrée, and a dessert. If the number of appetizers decreases from 7 to 5, find how many fewer possible three-course dinners the restaurant offers.

7. **RESTAURANT** Refer to Exercise 6. The manager increases the number of appetizers back to 7. Then she increases the number of desserts from 4 to 6. How many more possible three-course dinners are there than 168?

Get ConnectED *For more practice, go to* www.connected.mcgraw-hill.com.

Problem-Solving Practice

Problem-Solving Investigation: Act it Out

Solve each problem using any strategy you have learned.

1. **POLLS** Out of 200 people, 32% said that their favorite animal was a cat and 44% said that their favorite animal was a dog. How many more people chose a dog than a cat?

2. **PEACHES** Roi is picking peaches; he needs a total of $3\frac{1}{2}$ bushels of peaches. If he has already picked 3 bushels, how many more bushels does he need to pick?

3. **BASEBALL** Thirty-two teams are playing in the championship. If a team is eliminated once it loses, how many total games will be played in the championship?

4. **GEOMETRY** Draw the next two figures in the sequence.

5. **POOL RENTAL** The table below shows how much Ford Middle School was charged to rent the pool for a party based on the number of hours it was rented. Predict the cost for 5 hours.

Number of Hours	Cost
1	$120
1.5	$180
2	$240
2.5	$300

6. **GEOMETRY** Alyssa drove 55 miles per hour for 4 hours. Use the formula $D = rt$, where D is the distance, r is the rate, and t is the time to determine how far Alyssa drove.

7. **SCHOOL ELECTIONS** How many ways can a president, vice president, secretary, and treasurer be elected from a choice of 6 students?

8. **SHOPPING** Morty bought skis. The skis cost $215 and he received $35 in change. With how much money did Morty pay?

Multi-Part Lesson 3

PART E

Homework Practice

Use Data to Predict

1. **PETS** Of the dog owners polled, 26% clean their dogs' kennels monthly. Predict how many dog owners out of 330 clean their dogs' kennels monthly.

2. **CELL PHONES** In a survey, 94% of teens said they have a cell phone. Predict how many teens out of 548 do *not* have a cell phone.

3. **BOOKS** The results of a survey asking teens their favorite type of book to read is shown.

Favorite Type of Book

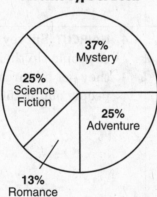

a. Out of 250 teens, predict how many would choose mystery as their favorite type of book to read.

b. Out of 250 teens, predict how many would choose adventure as their favorite type of book to read.

c. Out of 250 teens, predict how many would choose science fiction as their favorite type of book to read.

Get ConnectED *For more practice, go to* www.connected.mcgraw-hill.com.

Problem-Solving Practice

Use Data to Predict

---

1. **SPORTS** The circle graph shows the results of a poll to which 85 children aged 10–15 responded. If 1,020 children aged 10–15 are polled, how many would choose soccer as their favorite sport to watch?

Favorite Sport to Watch

- 8% Other
- 15% Soccer
- 12% Baseball
- 23% Tennis
- 42% Football

2. **SWIMMING POOL** The adults in a town were surveyed on whether they wanted a town swimming pool. The results are shown in the table below. Out of 580 adults, predict how many would say they wanted a town swimming pool.

Swimming Pool Survey	
Response	**Percent**
Yes	26
No	64
Undecided	10

---

3. **SWIMMING POOL** Use the table in Exercise 2 to predict how many adults out of 720 would be undecided about whether they wanted a town swimming pool.

4. **HAIRCUT** Survey results show that 68% of people tip their hairdresser when they get a haircut. Predict how many people out of 150 tip their hairdresser.

---

5. **GOLF** A survey showed that 28% of adults play golf in their free time. Out of 1,550 adults, predict how many would say they play golf.

6. **GOLF** Use the information in Exercise 5 to predict how many adults out of 1,550 would say they do *not* play golf.

Homework Practice

Unbiased and Biased Samples

Determine if each conclusion is valid. Justify your answer.

1. To determine the most common injury cared for in an emergency room, a reporter goes to the same hospital every afternoon for one month during the summer and observes people entering the emergency room. She concludes that second degree sunburn is the most common injury.

2. To evaluate customer satisfaction, a grocery store gives double coupons to anyone who completes a survey as they enter the store. The store manager determines that customers are very satisfied with their shopping experience in his store.

3. To evaluate the integrity of underground water lines, the department of public works randomly selects 20 sites in the city to unearth and observe the water lines. At 5 of the sites, the water lines needed repair. The department of public works concludes that about one-fourth of underground water lines throughout the city need repair.

4. **DOWNLOADS** A guidance counselor asked students who owned mobile phones, which was the last type of download each one downloaded to their mobile phone. The results are shown in the table. If there are 420 students in the school, how many can be expected to download ringtones?

Mobile Phone Downloads	
Type	**Frequency**
Games	10
Ringtones	25
Screensavers	14
Music	36

5. **DENTISTRY** A survey is to be conducted to determine the reasons dental patients are hesitant to go to the dentist. Describe the sample and explain why each sampling method might not be valid.
   a. Adults that are randomly selected from an office complex are asked to go online and fill out a questionnaire.

   b. A randomly selected dentist asks his patients why they may be hesitant to go to a dentist.

   c. Randomly selected dental patients who are having a routine check-up are asked to write down their feelings.

Get ConnectED *For more practice, go to* www.connected.mcgraw-hill.com.

Problem-Solving Practice

Unbiased and Biased Samples

FUNDRAISING For Exercises 1 and 2, use the survey results in the table at the right. Members of the Drama Club plan to sell popcorn as a fundraiser for their Shakespeare production. They survey 75 students at random about their favorite flavors of popcorn.

Flavor	Number
butter	33
cheese	15
caramel	27

1. What percent of the students prefer caramel popcorn?

2. If the club orders 400 boxes of popcorn to sell, how many boxes of caramel popcorn should they order? Explain how you found your answer.

DINING OUT For Exercises 3 and 4, use the following information. As people leave a restaurant one evening, 20 people are surveyed at random. Eight people say they usually order dessert when they eat out.

3. What percent of those surveyed say they usually order dessert when they eat out?

4. If 130 people have dinner at the restaurant tomorrow, how many would you expect to order dessert?

RECREATION For Exercises 5 and 6, use the table at the right which shows the responses of 50 people who expect to purchase a bicycle next year.

Bicycle Type	Number
mountain	11
touring	8
comfort	9
juvenile	19
other	3

5. What percent of those planning to buy a bicycle next year think they will buy a mountain bike?

6. If Mike's Bike Shop plans to order 1,200 bicycles to sell next year, how many mountain bikes should be ordered?

Homework Practice

Measures of Central Tendency

Find the mean, median, and mode for each set of data. Round to the nearest tenth if necessary.

1. number of shoppers at the store: 45, 14, 41, 45, 44, 64, 51

2. prices of towels: $10, $8, $20, $25, $14, $39, $10, $10, $8, $16

3. golf scores: −3, −2, +1, +1, −1, −1, +2, −6

4. points scored during football season: 14, 21, 3, 9, 19, 38, 21, 24, 56, 12, 7

5. Select the appropriate measure of central tendency to describe the data in the table. Justify your reasoning.

   Square Miles of Water by State

State	Water Area (mi²)
Alaska	91,316
Florida	11,828
Georgia	1,519
Minnesota	7,329
New Mexico	234
Virginia	3,180

6. Refer to the table of water area in Exercise 5. Which measure of central tendency is most affected by the extreme value?

7. **WORK** The table shows the hours Tabitha worked each week during the summer. How many hours did she work during the twelfth week to average 20 hours per week?

   Hours Worked

18	24	20	19	15	21
20	17	18	22	16	?

Get ConnectED *For more practice, go to* www.connected.mcgraw-hill.com.

Problem-Solving Practice

Measures of Central Tendency

1. **FOOTBALL** The table shows the number of games won by various teams in the NFL in a recent year. Find the mean, median, and mode of the data set.

Games Won by NFL Teams

Team	Games Won
Atlanta	4
Carolina	7
Denver	7
Green Bay	13
Jacksonville	11
Miami	1
New England	16
New Orleans	7
Oakland	4
San Francisco	5
Seattle	10
Tampa Bay	9

2. **FOOTBALL** Use the table in Exercise 1. Which measure of central tendency is most representative of the original set of data? Explain.

3. **PENCILS** The table shows the number of times per day that students sharpen their pencils. Which measure of central tendency is most affected by the extreme value?

Times Students Sharpen Pencils

2	3	0	1	2	2	3	4
0	5	2	5	2	5	2	4
2	4	6	0	5	6	5	5
2	0	0	1	4	6	10	2

4. **PENCILS** Use the table for Exercise 3. Would the mean, median, or mode best represent the original set of data? Explain.

Multi-Part Lesson 2

PART A

Homework Practice

Measures of Variation

1. **BLACK BEARS** Use the data in the table.

Weights of Black Bears (lb)
277 448 279 334 132 599 237 251 183 191

a. Find the range of the data.

b. Find the median and the upper and lower quartiles.

c. Find the interquartile range.

d. Find any outliers in the data.

2. **PRECIPITATION** Use the data of average monthly precipitation in Johnstown shown in the table.

Monthly Precipitation

Month	Jan.	Feb.	Mar.	Apr.	May.	Jun.	Jul.	Aug.	Sept.	Oct.	Nov.	Dec.
Inches	1.71	1.49	1.92	1.93	3.56	9.89	7.34	8.62	8.23	3.80	1.89	1.72

a. Find the range of the data.

b. Find the median and the upper and lower quartiles.

c. Find the interquartile range.

d. Find any outliers in the data and name them.

3. **TRAIN** The table shows the number of riders on the train each day for two weeks. Compare and contrast the measures of variation for both weeks.

Number of Riders on the Train		
Day	Week 1	Week 2
Monday	72	79
Tuesday	84	86
Wednesday	78	75
Thursday	67	49
Friday	86	137

Get ConnectED *For more practice, go to* www.connected.mcgraw-hill.com.

Problem-Solving Practice

Measures of Variation

FOOTBALL Use the table below that shows the winning scores in the Super Bowl.

Winning Super Bowl Scores, 1997–2008											
1997	1998	1999	2000	2001	2002	2003	2004	2005	2006	2007	2008
35	31	34	23	34	20	48	32	24	21	29	17

1. Explain how to find the range of the data. Then find the range.

2. Find the median, the upper and lower quartiles, and the interquartile range of the winning scores.

3. Describe how to find the limits for outliers. Then find the limits.

4. Are there any outliers among the winning Super Bowl scores? If so, what are they? Explain your reasoning.

GRADES Use the table showing the scores on a U.S. History test.

Scores on a U.S. History Test					
84	86	79	97	88	89
94	89	81	90	82	61
91	83	95	80	97	78

5. Find the range, median, upper and lower quartiles, and the interquartile range of the test scores.

6. Are there any outliers in this data? Explain your reasoning.

Homework Practice

Box-and-Whisker Plots

Draw a box-and-whisker plot for each set of data.

1. ages of children taking dance classes: 10, 8, 9, 7, 10, 12, 14, 14, 10, 16

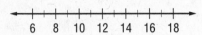

2. prices, in dollars, of bicycles: 150, 134, 136, 120, 145, 170, 125, 130, 145, 190, 140

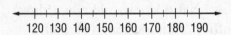

3. PRODUCTS Use the box-and-whisker plot that shows the average prices in cents per pound farmers received for eggs and wool.

Prices per pound received (¢)

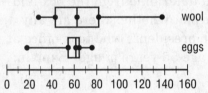

a. How do the median egg prices and the median wool prices compare?

b. How do the range in egg prices and the range in wool prices compare?

c. In the wool prices, which quartile shows the greatest spread of data?

d. About what percent of the data for the wool prices is above the upper quartile for the egg prices?

e. In general, do farmers get higher prices for eggs or wool? Justify your reasoning.

Get ConnectED *For more practice, go to* www.connected.mcgraw-hill.com.

Multi-Part
Lesson 2

PART B

Problem-Solving Practice

Box-and-Whisker Plots

U.S. VICE PRESIDENTS Use the box-and-whisker plot that shows the ages of U.S. vice presidents when they took office.

Ages of U.S. Vice Presidents

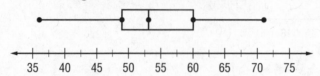

1. Describe the distribution of the data. What can you say about the ages of U.S. vice presidents?

2. What percent of U.S. vice presidents were at least 60 years old when they took office? Explain how you found your answer.

3. What percent of U.S. vice presidents were between the ages of 49 and 60 when they took office? Explain how you found your answer.

4. Can you determine from the box-and-whisker plot whether there are any U.S. vice presidents who took office at exactly age 55 years of age? Explain.

FIELD HOCKEY Use the box-and-whisker plot that shows the number of goals made by members of the field hockey team during the season.

Field Hockey Goals

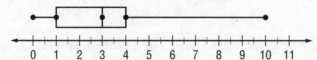

5. Describe the distribution of the data. What can you say about the number of goals made by the members of the field hockey team?

6. What percent of team members scored between 1 and 3 goals this season? Explain.

Multi-Part Lesson **3**

PART **B**

Homework Practice

Circle Graphs

Display each set of data in a circle graph.

1.

Volume of World's Oceans	
Ocean	Percent
Pacific	49%
Atlantic	26%
Indian	21%
Arctic	4%

2.

America's Energy Sources	
Type	Percent
Petroleum	40%
Natural Gas	23%
Coal	22%
Nuclear	8%
Other	7%

3. EXPORTS Use the circle graph that shows the percent of Persian Gulf petroleum exports by country.

  a. Which country has the most petroleum exports?

  b. How many times as much in exports does Iran have than Qatar?

Persian Gulf Exports

DISPLAYS **For each graph, find the value of *x*.**

4.   **Recycled Products**

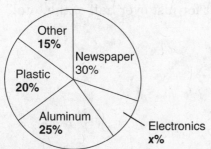

5.   **Time Management**

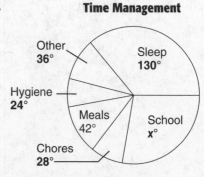

Get ConnectED *For more practice, go to* www.connected.mcgraw-hill.com.

Problem-Solving Practice

Circle Graphs

LANGUAGES Use the table that shows the number of people that speak the five languages that are spoken by the most people in the world.

| Languages Spoken by the Most People ||
Language	Speakers (millions)
Chinese, Mandarin	874
Hindi	366
English	341
Spanish	322
Bengali	207

1. Find the degrees for each part of a circle graph that shows the data.

2. Make a circle graph of the data. Which three languages account for 41% of the total?

MILITARY Use the table that shows the number of people active in the United States military in a recent year.

| United States Military, Active Duty ||
Branch	Personnel (thousands)
Army	538
Navy	333
Marine Corps	195
Air Force	328
Coast Guard	42

3. Make a circle graph of the data.

4. Which two branches taken together account for just over half of the total?

Multi-Part
Lesson **3**

PART C

Homework Practice

Histograms

1. **GOVERNMENT** The list gives the year of birth for each state governor in the United States in a recent year. Choose intervals and make a frequency table. Then construct a histogram to represent the data.

1943	1944	1956	1952	1970	1957	1950
1961	1964	1946	1942	1955	1941	1960
1942	1957	1953	1955	1948	1963	1951
1935	1946	1942	1963	1944	1940	1958
1948	1947	1956	1956	1957	1944	1947
1947	1956	1949	1954	1947	1942	1947
1950	1946	1966	1960	1959	1954	1950
1943						

2. **FOOTBALL** Use the histograms shown.

Scores of Winning Teams

a. Which bowl game had the higher winning team score?

b. In which bowl game was the winning team score in the interval 30–39 points more often?

c. Determine which bowl game has had a winning team score of at least 30 points more often.

d. What was the lowest winning team score in each bowl game? Explain.

Get ConnectED *For more practice, go to* www.connected.mcgraw-hill.com.

Problem-Solving Practice

Histograms

EXAMS For Exercises 1–3, use the histogram below that shows scores on a history test.

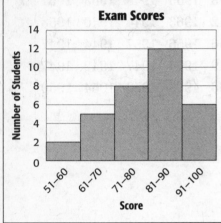

Exam Scores

MOVIES For Exercises 4–6, use the histogram below that shows movie revenues in a recent year.

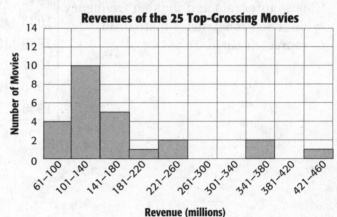

Revenues of the 25 Top-Grossing Movies

1. How many students scored at least 81 on the test? Explain how you found your answer.

2. How many students scored less than 81 on the exam? Explain how you found your answer.

3. Can you determine the highest grade from the histogram? Explain.

4. How many movies had a revenue of at least $141 million? Explain how you found your answer.

5. How many movies had a revenue of between $61 million and $180 million? Explain how you found your answer.

6. Can you determine how many movies had a revenue of between $121 and $140 million from the histogram? Explain.

Multi-Part Lesson 3

PART E

Homework Practice

Stem-and-Leaf Plots

Display each set of data in a stem-and-leaf plot.

1. {68, 63, 70, 59, 78, 64, 68, 73, 61, 66, 70}

2. {27, 32, 42, 31, 36, 37, 47, 23, 39, 31, 41, 38, 30, 34, 29, 42, 37}

3.

Major League Baseball Pitchers	
Player	**Wins**
J. Beckett	20
F. Carmona	19
J. Lackey	19
B. Webb	18
A. Harang	16
T. Hudson	16
K. Escobar	18
T. Wakefield	17
J. Peavy	19
J. Francis	17

4. Use the information in the stem-and-leaf plot shown below.

Pizzas Ordered by Convenience Store per Day

Stem	Leaf
1	2 3 9
2	4 4 7
3	1

$3 \mid 1 = 31$

a. What is the least number of pizzas ordered at the convenience store?

b. What is the range of the number of pizzas ordered?

c. Find the mode of the data.

d. Find the median of the data.

e. How many days are shown in the stem-and-leaf plot?

Get ConnectED *For more practice, go to* www.connected.mcgraw-hill.com.

Problem-Solving Practice

Stem-and-Leaf Plots

1. CUSTOMER SERVICE A restaurant owner recorded the average time in minutes customers waited to be seated each night. His data are shown in the table below. To organize the data into a stem-and-leaf plot, how many stems would you need?

Week 1	15	8	10	5	20	35	45
Week 2	9	3	7	8	25	38	43

2. PHONE Allison's mother makes a stem-and-leaf plot to track the time in minutes that Allison spends talking on the phone each night. In which interval are most of the Allison's calls?

Stem	Leaf
1	0 5
2	3 4 5 8 9
3	0 5 8
4	1 3 5

$1 \mid 5 = 15$ minutes

3. ELECTRIC BILLS Jenny's family is selling their house. Jenny's mother wants to put together a table of monthly electricity costs. Below is a list of their electric bills for the past twelve months. Display the data in a stem-and-leaf plot. In which interval are most of the electric bills?

$95, $99, $85, $79, $82, $88,

$98, $95, $94, $87, $89, $90

4. TEST SCORES The scores from the most recent test in Mr. James' biology class are shown in the stem-and-leaf plot below. Find the highest and lowest scores, and then write a statement that describes the data.

Stem	Leaf
5	4 5
6	3 7 8
7	0 1 5 5 8 9
8	0 2 3 7 9
9	0 3 5 8 8

$5 \mid 4 = 54\%$

5. SPORTS Use the following information.

Tamara recorded her times in seconds in the 100-meter dash from the past six track meets below.

16.7	16.4	16.1	17.0	16.5	16.8

a. Display the times in a stem-and-leaf plot.

b. What is Tamara's median time?

c. What is the difference between Tamara's best and worst times?

Course 2 • Statistical Displays

Multi-Part Lesson **4**
PART A

Homework Practice

Problem-Solving Investigation: Use A Graph

Mixed Problem Solving

For Exercises 1 and 2, use the table showing the price of loaves of bread.

Number of Loaves	Price ($)
1	3.50
2	7.00
3	10.50
4	14.00

1. Draw a graph of the data.

2. Use the graph in Exercise 1 to predict the price of 5 loaves of bread.

Use any strategy to solve Exercises 3–6.

3. LAWN TOOLS The bar graph shows the number of shovels and rakes sold during particular months at a hardware store. During which month was the number of rakes sold about twice the number of shovels sold?

4. NUMBER THEORY 42 is subtracted from 42% of a number. The result is 42. What is the number?

5. MONEY The value of the number of dimes is equal to the value of the number of quarters. If the total value of the quarters and dimes is $6.00, find the total number of coins.

6. SKIING Mrs. Roget is taking her family of 2 adults and 4 children skiing for the day. They need to rent ski equipment. What will it cost to ski for the day including equipment rental and lift tickets?

Daily Ski Costs		
Item	**Adults**	**Children**
Lift Ticket	$10.00	$8.00
Skis	$7.00	$4.25
Boots	$6.25	$4.25
Poles	$2.25	$1.75

Get ConnectED *For more practice, go to* www.connected.mcgraw-hill.com.

Course 2 • Statistical Displays

145

Multi-Part Lesson **4** PART A

Problem-Solving Practice

Problem-Solving Investigation: Use A Graph

For Exercises 1–3, use the table showing the area covered by boxes of tiles.

Number of Boxes of Tiles	Area (ft²)
1	15
2	30
3	45

1. Draw a graph of the data.

2. Predict the number of boxes that would cover 60 square feet.

3. Predict the square feet that will be covered by 6 boxes of tiles.

4. EXERCISE Sherman wants to begin a new exercise program. His goal is to begin by exercising for 25 minutes. He goes to the gym two times a week, increasing his workout by five minutes each time. How long will it take him to work up to an hour?

5. MONEY Brianna made a $13.82 purchase at the grocery store. She received two bills and five coins in change. What denomination of bill did she pay with?

6. Refer to Exercise 5. What bills and coins did she receive as change?

7. NUMBER THEORY A number is multiplied by 32 then divided by 14. The square root of the result is 4. What is the number?

8. PIZZA Russell has his choice of five pizza toppings: onions, sausage, mushrooms, pepperoni, and green pepper. In order to get a special price, he can only choose two toppings. How many combinations of toppings could he choose?

Homework Practice

Scatter Plots and Lines of Best Fit

1. **WATER LEVEL** Use the graph that shows the level of rising water of a lake after several days of rainy weather.

   a. If the water continues to rise, predict the day when the water level will be above flood stage of 20.5 feet.

   b. How many days did it take for the water level to rise 4 feet?

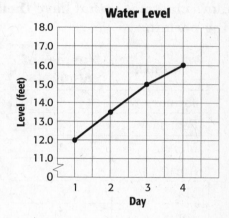

2. **PROPERTY** Use the table that shows the property value per acre for five years.

   a. Make a scatter plot of the data. Use the time on the horizontal axis and the property value on the vertical axis.

Property Value (per acre)	
Time	Value
2007	$14,000
2008	$16,600
2009	$18,900
2010	$21,500
2011	$24,000

   b. Describe the relationship, if any, between the two sets of data.

   c. Predict the property value per acre in 2012.

Get ConnectED For more practice, go to www.connected.mcgraw-hill.com.

Problem-Solving Practice

Scatter Plots and Lines of Best Fit

For Exercises 1–3, use the table that shows the relationship between the month of the year and the number of Tamika's classmates that have their driving permit.

Month	Number of Students
January	1
February	3
March	4
April	5
May	8
June	10
July	11
August	14
September	15
October	15
November	18
December	21

1. Make a scatter plot of the data. Label the horizontal axis, Month; and the vertical axis, Number of Students.

2. Describe the relationship, if any, between the two sets of data.

3. Why do you think this relationship exists?

For Exercises 4–6, use the graph that shows the time it takes Vernon to complete a 26-mile marathon.

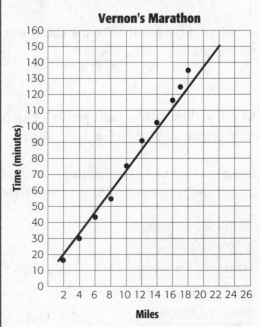

4. Predict the time it will take Vernon to reach Mile 22 of the marathon and how long it will take Vernon to complete the marathon.

5. For how many minutes will he have run when he reaches the 8-mile mark?

6. How many miles will he have run in 140 minutes?

Homework Practice

Select an Appropriate Display

Select an appropriate type of display for each situation. Justify your reasoning.

1. the numbers of students who spend Sundays doing homework, visiting with friends, and/or working

2. the number of each of four types of flowers found in a garden

3. prices of scuba gear in a store arranged by intervals

4. the spread of times for boaters completing a yacht race

Select an appropriate type of display for each situation. Justify your reasoning. Then construct the display. What can you conclude from your display?

5.

Number of Miles of Gulf Coastline	
State	**Miles**
Alabama	53
Florida	770
Louisiana	397
Mississippi	44
Texas	367

6. **FLUTE** Tobey practiced the flute for 15 minutes on Monday. On Tuesday, he practiced 15 more minutes than he did on Monday. On Wednesday, he practiced 15 more minutes than he did on Tuesday. The pattern continued through Saturday.

Get ConnectED *For more practice, go to* www.connected.mcgraw-hill.com.

Multi-Part Lesson 4

PART C

Problem-Solving Practice

Select an Appropriate Display

AGE For Exercises 1 and 2, use the following information. The table shows the ages of people at a roller-skating rink.

Ages of People Roller Skating	
Age	**Number of People**
10 and under	19
11–20	22
21–30	14
31–40	7
over 40	6

1. Select an appropriate display for the data. Justify your reasoning.

2. Construct the display.

3. VEGETABLES A survey asked students which vegetable they prefer. Of those who responded, 17 said corn, 22 said carrots, 9 said both corn and carrots, and 7 said neither vegetable. Select an appropriate display for this data.

4. Construct the display in Exercise 3.

5. TELEVISIONS The table shows the number of televisions that were sold. Select an appropriate display for this data.

Television Sales by Screen Size	
Size (in.)	**Percent**
20	10
27	39
42	36
46	15

6. Construct the display in Exercise 5.

Homework Practice

Volume of Prisms

Find the volume of each prism. Round to the nearest tenth if necessary.

1.

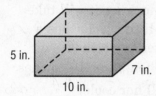

5 in.
7 in.
10 in.

2.

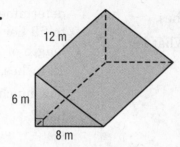

12 m
6 m
8 m

3.

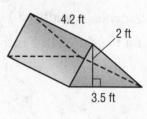

4.2 ft
2 ft
3.5 ft

4.

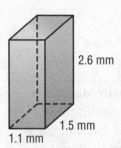

2.6 mm
1.5 mm
1.1 mm

5.

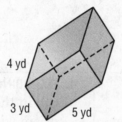

4 yd
3 yd
5 yd

6.

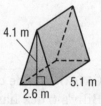

4.1 m
2.6 m
5.1 m

ESTIMATION Estimate to find the approximate volume of each prism.

7.

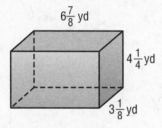

$6\frac{7}{8}$ yd
$4\frac{1}{4}$ yd
$3\frac{1}{8}$ yd

8.

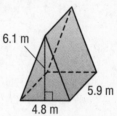

6.1 m
4.8 m
5.9 m

9. MAIL The United States Post Office has two different priority mail flat rate boxes. Which box has the greater volume? Justify your answer.

Box 1: $6\frac{1}{2}$ in. × $8\frac{1}{2}$ in. × 11 in. Box 2: $3\frac{3}{8}$ in. × $11\frac{7}{8}$ in. × $13\frac{5}{8}$ in.

Get ConnectED *For more practice, go to* <u>www.connected.mcgraw-hill.com</u>.

Problem-Solving Practice

Volume of Prisms

---

1. **PACKAGING** A cereal box has a length of 8 inches, a width of $1\frac{3}{4}$ inches, and a height of $12\frac{1}{8}$ inches. What is the volume of the cereal box?

2. **FOOD STORAGE** Nara wants to determine how much ice it will take to fill her cooler. If the cooler has a length of 22 inches, a width of 12 inches, and a height of $10\frac{1}{2}$ inches, how much ice will her cooler hold?

3. **TRANSPORTATION** The cargo-carrying part of Billy's truck has a length of 8.3 meters, a width of 3 meters, and a height of 4.2 meters. What is the maximum volume of sand that Billy's truck can carry?

4. **PLUMBING** Alexia's bathroom has a tub in the shape of a rectangular prism with a length of 1.5 meters, a width of 0.5 meter, and a height of 0.4 meter. How many cubic meters of water can it hold?

5. **STICKY NOTES** A triangular box of sticky notes is shown. Find the volume of the box.

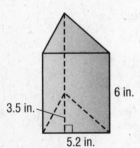

6. **GEOMETRY** A *pentagonal prism* is a prism that has bases that are pentagons. Use $V = Bh$ where B is the area of the base, to find the volume of the pentagonal prism below.

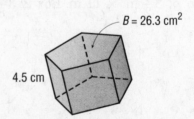

$B = 26.3 \text{ cm}^2$

4.5 cm

Homework Practice

Volume of Cylinders

Find the volume of each cylinder. Round to the nearest tenth.

1.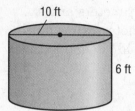
10 ft
6 ft

2.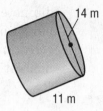
14 m
11 m

3.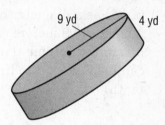
9 yd   4 yd

4.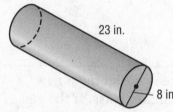
23 in.
8 in.

5.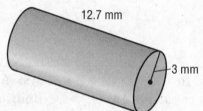
12.7 mm
3 mm

6.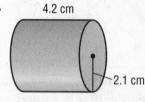
4.2 cm
2.1 cm

7. CONTAINER What is the volume of a cylindrical barrel that has a diameter of $1\frac{1}{2}$ feet and a height of 4 feet?

ESTIMATION Match each cylinder with its approximate volume.

8. diameter = 4 cm, height = 3.6 cm   **a.** 116 cm³

9. radius = 2.7 cm, height = 5 cm   **b.** 115 cm³

10. radius = 3 cm, height = 4.1 cm   **c.** 106 cm³

11. diameter = 8.2 cm, height = 2 cm   **d.** 45 cm³

12. FUEL Two fuel tanks with the dimensions shown have the same volume. What is the value of h?

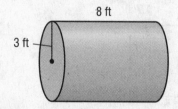
8 ft
3 ft

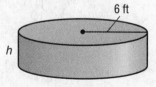
6 ft
h

Get ConnectED *For more practice, go to* www.connected.mcgraw-hill.com.

Problem-Solving Practice
Volume of Cylinders

1. **WATER STORAGE** A cylindrical water tank has a diameter of 5.3 meters and a height of 9 meters. What is the maximum volume that the water tank can hold? Round to the nearest tenth.

2. **PACKAGING** A can of corn has a diameter of 6.6 centimeters and a height of 9.9 centimeters. How much corn can the can hold? Round to the nearest tenth.

3. **CONTAINERS** Felisa wants to determine the maximum capacity of a cylindrical bucket that has a radius of 6 inches and a height of 12 inches. What is the capacity of Felisa's bucket? Round to the nearest tenth.

4. **GLASS** Antoine is designing a new, cylindrical drinking glass. If the glass has a diameter of 8 centimeters and a height of 12.8 centimeters, what is its volume? Round to the nearest tenth.

5. **PAINT** A can of paint is 15 centimeters high and has a diameter of 13.6 centimeters. What is the volume of the can? Round to the nearest tenth.

6. **SPICES** A spice manufacturer uses a cylindrical dispenser like the one shown. Find the volume of the dispenser to the nearest tenth.

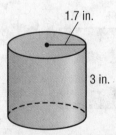

1.7 in.

3 in.

Homework Practice

Volume of Pyramids

Find the volume of each pyramid. Round to the nearest tenth if necessary.

1.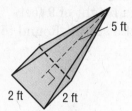
   5 ft
   2 ft   2 ft

2.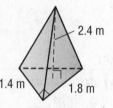
   2.4 m
   1.4 m   1.8 m

3.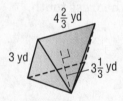
   $4\frac{2}{3}$ yd
   3 yd $3\frac{1}{3}$ yd

4.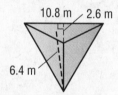
   10.8 m   2.6 m
   6.4 m

Find the height of each pyramid.

5. square base edge 15 feet and volume 1,350 cubic feet

6. triangular base with base edge 12 inches, base height 9 inches, and volume 108 cubic inches

7. **GREAT PYRAMID** The Great Pyramid has a height of about 480.7 feet and base edges about 756 feet. The base is almost square. Find the approximate volume of this pyramid.

Get ConnectED *For more practice, go to* www.connected.mcgraw-hill.com.

Multi-Part Lesson 1

PART E

Problem-Solving Practice

Volume of Pyramids

1. **SOUVENIRS** On a trip to Oregon, Sabrina bought a small stone in the shape of a square pyramid as a souvenir. Find the volume of the stone. Round to the nearest tenth.

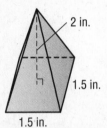

2. **ART** An artist created a statue in the shape of a triangular pyramid. The triangular base has a height of 9 feet. Find the volume of the statue. Round to the nearest tenth.

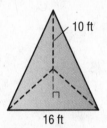

3. **GATE POST** The top of a gate post is in the shape of a square pyramid. The height of the pyramid is 5 inches and each side of the base is 7.4 inches. Find the volume of wood needed to make the top of the gate post. Round to the nearest tenth.

4. **DISPLAY STAND** A glass stand to display a doll is in the shape of a right triangular pyramid as shown. Find the volume. Round to the nearest tenth.

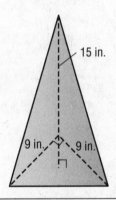

5. **COAL** A piece of coal is in the shape of a square pyramid. Find the volume. Round to the nearest tenth.

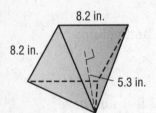

6. **ART PROJECT** An art class builds a square pyramid with sides 12 foot wide. The pyramid is 17 feet high. Each student in the school deposits a colored cube with side length of 1 foot into the pyramid. To the nearest hundred, about how many students are in the school?

Course 2 • Volume and Surface Area

Multi-Part
Lesson 1
PART F

Homework Practice

Volume of Cones

Find the volume of each cone. Round to the nearest tenth.

1.

2 in.

5 in.

2.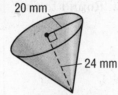

20 mm

24 mm

3.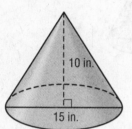

10 in.

15 in.

4.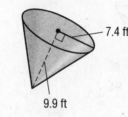

7.4 ft

9.9 ft

5.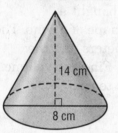

14 cm

8 cm

6.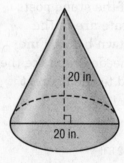

20 in.

20 in.

7.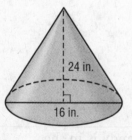

24 in.

16 in.

8.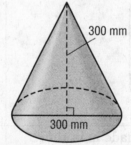

300 mm

300 mm

9. height: 26.8 centimeters; radius: 12 centimeters

10. height: 34 feet; diameter: 9.8 feet

Find the area of the base of each cone.

11. volume: 36 cubic inches; height: 9 inches

12. volume: 238 cubic centimeters; height: 74 centimeters

Get ConnectED *For more practice, go to* www.connected.mcgraw-hill.com.

Multi-Part
Lesson **1**

PART **F**

Problem-Solving Practice

Volume of Cones

1. DESSERT Find the volume of the ice cream cone shown below. Round to the nearest tenth.

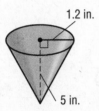

1.2 in.

5 in.

2. SALT Lecretia uses a small funnel as shown below to fill her salt shaker. Find the volume of the funnel. Round to the nearest tenth.

2 in.

0.5 in.

3. ENTRYWAY The top of the stone posts at the entry to an estate are in the shape of a cone as shown below. Find the volume of stone needed to make the top of the post. Round to the nearest tenth.

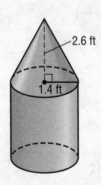

2.6 ft

1.4 ft

4. PAPERWEIGHT Marta bought a paperweight in the shape of a cone. The radius was 10 centimeters and the height 9 centimeters. Find the volume. Round to the nearest tenth.

5. LAMPSHADE A lampshade is in the shape of a cone. The diameter is 5 inches and the height 6.5 inches. Find the volume. Round to the nearest tenth.

6. CANDY A piece of candy is in the shape of a cone. The height of the candy is 2 centimeters and the diameter is 1 centimeter. Find the volume. Round to the nearest tenth.

Multi-Part
Lesson **2**

PART **B**

Homework Practice

Surface Area of Prisms

Find the surface area of each prism. Round to the nearest tenth if necessary.

1.
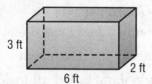
3 ft
6 ft
2 ft

2.
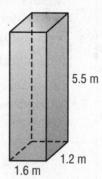
5.5 m
1.6 m
1.2 m

3.
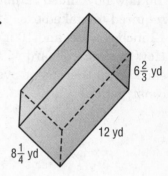
$6\frac{2}{3}$ yd
12 yd
$8\frac{1}{4}$ yd

4.
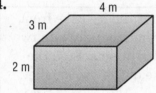
4 m
3 m
2 m

5.
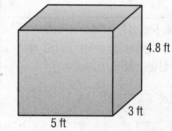
4.8 ft
5 ft
3 ft

6.
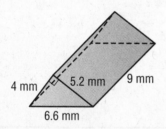
4 mm
5.2 mm
9 mm
6.6 mm

7.
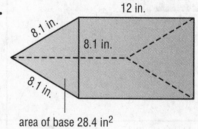
12 in.
8.1 in.
8.1 in.
8.1 in.
area of base 28.4 in²

8. BIRTHDAY GIFT When wrapping a birthday gift for his mother, Kenji adds an additional 2.5 square feet of gift wrap to allow for overlap. How many square feet of gift wrap will Kenji use to wrap a gift 3.5 feet long, 18 inches wide, and 2 feet high?

9. CONTAINERS A company needs to package hazardous chemicals in special plastic rectangular prism containers that hold 80 cubic feet. Find the whole number dimensions of the container that would use the least amount of plastic.

Get ConnectED *For more practice, go to* www.connected.mcgraw-hill.com.

Problem-Solving Practice
Surface Area of Prisms

1. **PACKAGING** A packaging company needs to know how much cardboard will be required to make boxes 18 inches long, 12 inches wide, and 10 inches high. How much cardboard will be needed for each box if there is no overlap in the construction?

2. **INSULATION** Jane needs to buy insulation for the inside of a truck container. The container is a rectangular prism 15 feet long, 8 feet wide, and $7\frac{1}{2}$ feet high. How much insulation should Jane buy if all inside surfaces except the floor are to be insulated?

3. **ICE** Suppose the length of each edge of a cube of ice is 4 centimeters. Find the surface area of the cube.

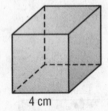

4 cm

4. **ICE** Suppose you cut the ice cube from Exercise 3 in half horizontally into two smaller rectangular prisms. Find the surface area of one of the two smaller prisms.

5. **CONTAINERS** What is the total surface area of the inside and outside of a container in the shape of an equilateral triangular prism with a triangular base 5 meters on each side, area of base 10.8 square meters, and height of prism 2.2 meters? Round to the nearest tenth.

6. **TOYS** Oscar is making a play block for his baby sister by gluing fabric over the entire surface of a foam block. How much fabric will Oscar need?

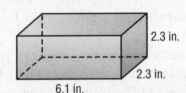

2.3 in.

2.3 in.

6.1 in.

Homework Practice

Surface Area of Cylinders

Find the surface area of each cylinder. Round to the nearest tenth.

1. 4 in.   15 in.

2. 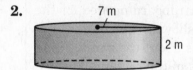 7 m

2 m

3.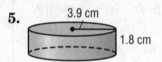
26 cm

12 cm

4. 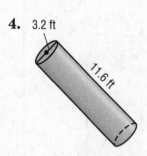 3.2 ft

11.6 ft

Estimate the surface area of each cylinder.

5. 3.9 cm

1.8 cm

6. 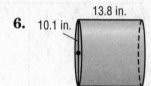 13.8 in.

10.1 in.

7. FENCE POST A cylindrical wooden fence post has a radius of 4 inches and a height of 48 inches. Find the surface area of the fence post. Round to the nearest tenth.

8. POSTER Walt is wrapping a poster enclosed in a cylindrical tube. The tube has a diameter of 6 centimeters and a length of 50 centimeters. Find the amount of wrapping paper Walt needs. Round to the nearest tenth.

Get ConnectED *For more practice, go to www.connected.mcgraw-hill.com.*

Problem-Solving Practice

Surface Area of Cylinders

1. **CONTAINERS** A company is comparing the amount of cardboard needed for the two containers shown. The volume of the containers is about the same. Find the surface area of the rectangular prism container. Round to the nearest tenth.

4.4 in.

10 in.

10 in.

3.8 in.

4 in.

2. **CONTAINERS** Find the surface area of the cylindrical container in Exercise 1. Round to the nearest tenth. Which container has the smaller surface area?

3. **PILLOW** Brooke is making a cylindrical bolster pillow for her couch. The pillow is 18 inches long and has a radius of 5 inches. Find the amount of fabric needed for the pillow. Round to the nearest tenth.

4. **STAIRWELL** The stairwell in a museum is in the shape of a cylinder. The diameter is 20 feet and the height is 100 feet. The museum needs to paint the inside of the stairwell. Find the surface area. Round to the nearest tenth.

5. **WATER SLIDE** A tube for a water slide is cylindrical. The tube is 40 yards long and has a radius of 2 yards. Find the surface area. Do not include the open ends. Round to the nearest tenth.

6. **TUNNEL** A tunnel over a highway is in the shape of half a cylinder as shown. It is open at both ends. Find the surface area of the inside of the tunnel. Round to the nearest tenth. Do not include the bottom which is the highway.

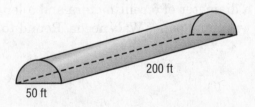

200 ft

50 ft

Multi-Part Lesson **2**

PART **E**

Homework Practice

Surface Area of Pyramids

Find the surface area of each pyramid. Round to the nearest tenth if necessary.

1.

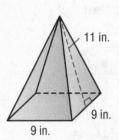

11 in.

9 in.

9 in.

2.

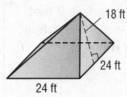

18 ft

24 ft

24 ft

3.

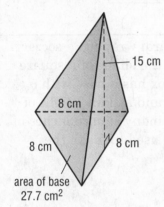

15 cm

8 cm

8 cm   8 cm

area of base
27.7 cm²

4.

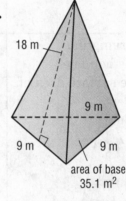

18 m

9 m

9 m   9 m

area of base
35.1 m²

5.

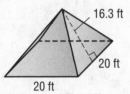

16.3 ft

20 ft

20 ft

6.

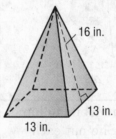

16 in.

13 in.

13 in.

7.

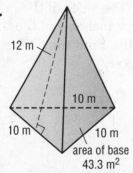

12 m

10 m

10 m   10 m

area of base
43.3 m²

8.

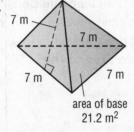

7 m

7 m

7 m   7 m

area of base
21.2 m²

9. MODEL HOUSE Baron built a square pyramid block to use as the roof of a model house he was making. The square base had sides of length 8 inches and the slant height was 6 inches. Find the surface area of the block. Round to the nearest tenth.

Get ConnectED *For more practice, go to* www.connected.mcgraw-hill.com.

Problem-Solving Practice

Surface Area of Pyramids

1. **PORCH** Lucille has a screened porch in the shape of a square prism. The roof is a square pyramid. If the roof is 9 feet by 9 feet and the slant height is 6 feet, find the lateral area of the roof.

2. **TENT** The Summers children are camping out in the tent shown. Find the lateral area of the tent.

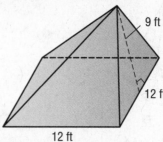

9 ft

12 ft

12 ft

3. **BOX** Alexander has a triangular pyramid box to keep his holiday decorations in. Find the surface area of the box.

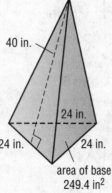

40 in.

24 in.

24 in.   24 in.

area of base
249.4 in²

4. **SOCCER BALL** Sarai wrapped a soccer ball in a box in the shape of a square pyramid. The box has a base with sides 12 inches and a slant height of 20 inches. How many square inches of cardboard were used to make the box?

5. **MOBILE** Gemma made a mobile to hang over her brother's crib. She put each animal on the mobile in a clear plastic case. The shape of one case was an equilateral triangular pyramid as shown. Find the surface area of the pyramid.

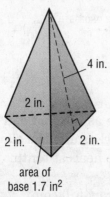

4 in.

2 in.

2 in.   2 in.

area of
base 1.7 in²

6. **MOBILE** Another clear plastic case on the mobile in Exercise 5 is in the shape of a square pyramid. The sides of the square base are each 2 inches and the slant height is 4 inches. Find the surface area of this case.

Multi-Part Lesson 3

PART A

Homework Practice

Problem-Solving Investigation: Solve a Simpler Problem

Mixed Problem Solving

Solve Exercises 1 and 2. Use the *solve a simpler problem* strategy.

1. **STADIUM** The exits in a stadium are designed to allow 1,200 people to leave the stadium each minute. At this rate, how long would it take for 10,800 people to leave the stadium?

2. **PHARMACY** A city has three major pharmacy chains which have a total of 895,000 customers. Approximately how many customers do business at each major pharmacy?

Pharmacy	Percent
A	54.8
B	32.4
C	12.8

Use any strategy to solve Exercises 3–7.

3. **CARPENTRY** Mr. Fernandez uses 7 boards that are 4 feet long and 6 inches wide to make one bookshelf. If he buys lumber in lengths of 8 feet with a width of 12 inches, how many pieces of lumber does he need to purchase to make 5 bookshelves?

4. **AREA** Rosie is making a stained glass window above her front doorway in the shape as shown in the figure. To the nearest tenth, what is the area of the shaded portion of the window?

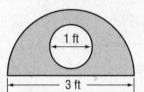

5. **QUALITY CONTROL** For every 250 televisions tested, 3 televisions are found to be defective. How many televisions were tested if 48 televisions were found defective?

6. **APPLIANCE REPAIR** An appliance repair company charged $35 to make a house call. After arriving, the company charged $10 for every 15 minutes of labor. How much was the repair bill if the new parts cost $23 and the appliance took 45 minutes to repair?

7. **FOAM BLOCKS** A small chunk of foam has been removed from the foam cube. What is the volume of the remaining foam?

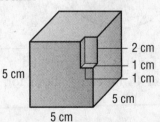

Get ConnectED *For more practice, go to* www.connected.mcgraw-hill.com.

Problem-Solving Practice

Problem-Solving Investigation: Solve a Simpler Problem

Solve each problem using any strategy you have learned.

1. **AREA** Find the area of the figure below. Round to the nearest tenth.

   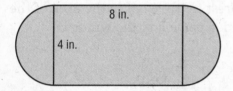

2. **MONEY** The table below shows the amount of money Shoshi earned for working various hours. Write a rule to represent the amount of pay P based on the number of hours worked h.

Hours	1	2	3
Pay	5.50	11.00	16.50

3. **SALES** For every nickel increase in price, the subscriptions to the Perrysville Paper decreases by 5 people. If 1,256 people currently subscribe to the paper, how many people will subscribe to it if the price is increased by $0.25?

4. **SCALE DRAWING** Shannon is creating a scale drawing of her classroom. If she is using the scale $\frac{1}{2}$ inch = 1 foot and the room model is 10 inches by 15 inches, what are the dimensions of the actual room?

5. **STUDY TIME** The circle graph below shows the results to a survey asking students how long they study each night. In a school of 400 students, how many students study 1.5–2.5 hours per night?

   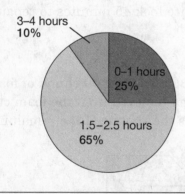

6. **PHOTOGRAPHY** What is the area of the matting around the picture below?

   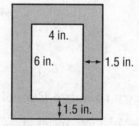

7. **TRAVEL** How far has Kim traveled if she has driven 45 miles per hour for 4 hours?

8. **SISTERS** Jolene is 3 years older than Susie. Cyd is 2 years younger than Susie. If Cyd is 10 years old, how old are Susie and Jolene?

Multi-Part Lesson 3

PART C

Homework Practice

Volume and Surface Area of Composite Figures

Find the volume of each composite figure. Round to the nearest tenth.

1.
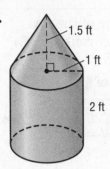
1.5 ft
1 ft
2 ft

2.
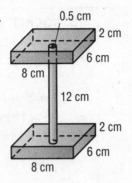
0.5 cm
2 cm
6 cm
8 cm
12 cm
2 cm
6 cm
8 cm

3.
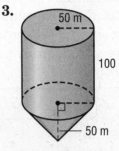
50 m
100
50 m

4. **TABLE** Rina is building a table as shown in the figure. Find the volume of wood she needs for the table. The cylindrical table legs are 24 inches tall with a diameter of 2 inches.

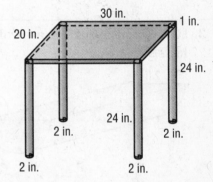
30 in.
20 in.
1 in.
24 in.
24 in.
2 in.
2 in.
2 in.
2 in.

Find the surface area of each composite shape. Round to the nearest tenth if necessary.

5.
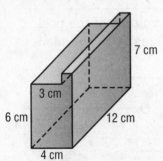
7 cm
3 cm
6 cm
12 cm
4 cm

6.
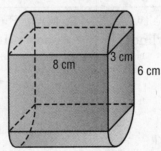
8 cm
3 cm
6 cm

7.
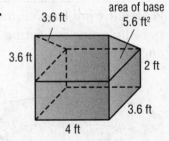
area of base
5.6 ft²
3.6 ft
3.6 ft
3.6 ft
2 ft
4 ft
3.6 ft

Get ConnectED *For more practice, go to* www.connected.mcgraw-hill.com.

Problem-Solving Practice

Volume and Surface Area of Composite Figures

Solve each problem using any strategy you have learned. Round to the nearest tenth if necessary.

1. LUNCH BOX Find the volume of the lunch box shown. The container consists of a rectangular prism and a triangular prism.

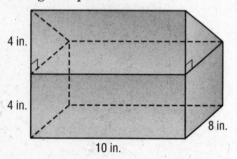

2. ROCKET Find the volume of the toy rocket shown. The rocket consists of a cylinder and a cone.

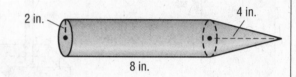

3. LOG Find the surface area of the half-log shown.

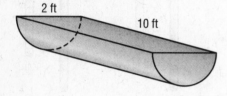

4. LOG Find the volume of the log in Exercise 3.

5. PLANTER Find the volume of the two-tiered planter consisting of a cylinder and a cone.

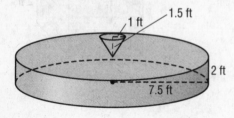

6. POOL Find the volume of the pool shown. It consists of two cylinders.

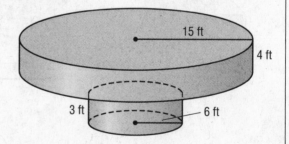

Multi-Part Lesson 1

PART B

Homework Practice

Convert Customary Units

Complete.

1. 4 c = _____ fl oz

2. 5 c = _____ pt

3. 3 lb = _____ oz

4. 24 ft = _____ yd

5. $1\frac{1}{2}$ pt = _____ c

6. 64 oz = _____ lb

7. 4 mi = _____ ft

8. $2\frac{3}{4}$ mi = _____ ft

9. 3,000 lb = _____ T

10. 5 gal = _____ qt

11. $3\frac{1}{4}$ qt = _____ pt

12. $4\frac{5}{8}$ T = _____ lb

13. $3\frac{1}{2}$ gal = _____ qt

14. 7 c = _____ qt

15. 40 fl oz = _____ qt

16. 660 yd = _____ mi

17. 1.9 yd = _____ in.

18. $2\frac{1}{4}$ T = _____ oz

19. SPORTS The track surrounding a football field is $\frac{1}{4}$ mile long. How many yards long is the track?

20. STRAWBERRIES One quart of strawberries weighs about 2 pounds. How many quarts of strawberries would weigh $\frac{1}{4}$ ton?

21. ANALYZE GRAPHS Use the graph shown.

    a. What does an ordered pair from this graph represent?

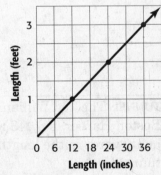

    b. Write two sentences that describe the graph.

    c. Explain how you could use the graph to find the length in inches of a 1.5 foot iguana.

Get ConnectED *For more practice, go to* www.connected.mcgraw-hill.com.

Course 2 • Measurement and Proportional Reasoning

Problem-Solving Practice

Convert Customary Units

---

1. WEIGHT The average weight of a baby at birth is 7 pounds. How many ounces is the average weight of a baby?

2. WATERFALLS The height of Niagara Falls is 182 feet. How many yards high is Niagara Falls?

---

3. GASOLINE The gasoline tank of a minivan holds 18 gallons. How many quarts can the tank hold?

4. CELL PHONES Cell phones can weigh as little as 2 ounces. How many pounds can the cell phone weigh?

---

5. RECIPE A recipe for ice cream calls for 56 fluid ounces of milk. How many cups of milk are in the recipe?

6. STATUE The Statue of Liberty weighs 450,000 pounds. How many tons does the statue weigh?

---

7. TUNNEL The Ted Williams Tunnel under Boston Harbor is 8,448 feet long. How many yards is the length of the tunnel?

8. COAL The United States exports over 200 billion pounds of coal. How many tons does the United States export?

Course 2 • Measurement and Proportional Reasoning

Homework Practice

Convert Metric Units

Complete.

1. 570 cm = _?_ m

2. 356 mm = _?_ m

3. 4.7 m = _?_ cm

4. 0.4 m = _?_ mm

5. 0.63 cm = _?_ mm

6. 0.18 mm = _?_ cm

7. 0.42 km = _?_ m

8. 0.09 km = _?_ m

9. 0.13 m = _?_ cm

10. 27 kg = _?_ g

11. 8.3 g = _?_ mg

12. 257 mg = _?_ g

13. 486 g = _?_ kg

14. 55.5 g = _?_ kg

15. 68,700 g = _?_ kg

16. 308 mL = _?_ L

17. 1.7 L = _?_ mL

18. 88 L = _?_ kL

19. 0.059 kL = _?_ L

20. 64,000 mL = _?_ L

21. 30,000 mL = _?_ L

Order each set of measures from least to greatest.

22. 0.06 km, 47 m, 15,800 cm

23. 891 g, 7,800 mg, 0.5 kg

24. CAVES The survey length of an underground cave is 0.914 kilometer. How many meters in length is this cave?

25. FOOD A 15-ounce box contains 0.425 kilogram of cereal. How many grams of cereal are in the box?

 *For more practice, go to www.connected.mcgraw-hill.com.*

Problem-Solving Practice

Convert Metric Units

---

1. RUNNING Each morning Arthur runs 1.5 kilometers. How many meters does he run each morning?

2. AVIATION A helicopter was flying 800 meters above the ground. How many kilometers above the ground was it flying?

---

3. SODA A soda can contains 355 milliliters of liquid. How many liters of liquid does it contain?

4. CONSTRUCTION The ceilings of most classrooms are about 2.5 meters above the floor. How many centimeters high is the ceiling?

---

5. FENCING Crystal's garden is 1,270 centimeters around the edges. How many meters of fencing material does she need to enclose her garden?

6. GARDENING Mr. Chou's lawn sprinker sprays about 150 liters of water each hour. How many kiloliters of water does it spray each hour?

---

7. NUTRITION For 11 to 14 year olds, the Recommended Dietary Allowance (RDA) for protein is about 60 grams daily. How many milligrams do they need daily?

8. PETS Amal's cat has a mass of about 4,000 grams. How many kilograms is the weight of her cat?

Homework Practice

Convert Between Systems

Complete. Round to the nearest hundredth if necessary.

1. 1.82 m ≈ _____ yd

2. 3.6 mi ≈ _____ km

3. 0.95 m ≈ _____ ft

4. 6.8 yd ≈ _____ m

5. 3.4 qt ≈ _____ mL

6. 825 mL ≈ _____ pt

7. 8.41 L ≈ _____ gal

8. 14.3 lb ≈ _____ kg

9. 762.8 g ≈ _____ lb

10. 8.5 in. ≈ _____ cm

11. 94 cm ≈ _____ in.

12. 125 mL ≈ _____ c

13. 3 c ≈ _____ mL

14. 210 lb ≈ _____ kg

Determine which measurement is greater.

15. 2 yd, 1.7 m

16. 7 lb, 4 kg

17. 6 gal, 22 L

18. Order the following measures from least to greatest:
0.5 m, 30 in., 1.25 ft, 17 cm

19. CASSEROLE Cassandra used 2.8 pounds of ground beef in a recipe for hamburger casserole. About how many kilograms is the mass of the ground beef?

20. BEDROOM Hayfa measured the length of her room and found that it was 4.5 yards long. About how many meters is the length of her room?

For more practice, go to www.connected.mcgraw-hill.com.

Problem-Solving Practice

Convert Between Systems

1. RACE Leola ran in a 10-kilometer race. About how many miles did she run?

2. SUPPER Dallison cooked a 5-pound roast for supper. What is the estimated mass in gram?

3. SWIMMING POOL Nykia swam the length of her swimming pool twice. The dimensions of her pool are shown below. About how many meters did she swim?

7 ft

21 ft

4. LEMONADE Beryl made 5 gallons of lemonade for a family gathering. About how many liters of lemonade did Beryl make?

5. MATH BOOK The length of Yuan's math book is 10.5 inches. What is the approximate length of her book in centimeters?

6. YOGURT Leatrice bought two one-quart containers of frozen yogurt. About how many liters of frozen yogurt did she buy?

7. AUTOMOBILES Mr. Shelton's car weighs about 1.75 tons. Find the approximate mass of his car in kilograms.

8. ELEVATORS The elevator in a new building travels a maximum distance of 32 meters. Find the estimate distance traveled in yards.

Multi-Part Lesson 1

PART E

Homework Practice

Convert Rates

Complete. Round to the nearest tenth if necessary.

1. 345 ft/min = _____ ft/h

2. 64 mi/h ≐ _____ ft/s

3. 17 cm/min = _____ m/h

4. 815 gal/h ≈ _____ L/sec

5. 39 ft/min ≈ _____ cm/s

6. 6,000 kg/day ≈ _____ T/wk

7. 110 mi/h = _____ mi/day

8. 2 lb/wk ≈ _____ kg/day

9. 90 ft/h ≈ _____ m/h

10. 44 mi/h ≈ _____ km/min

11. 22 lb/day = _____ oz/h

12. 720 pt/h = _____ qt/min

Use the table below showing the speed in miles per hour of several bikers.

Speeds While Biking	
Name	**Speed (miles per hour)**
Zoe	24
Jermaine	22.5
Ragsak	31.8
Yuzo	27

13. What is Ragsak's speed in feet per second? Round to the nearest tenth.

14. What is Zoe's estimated speed in meters per minute?

15. How many feet per minute faster is Yuzo's speed than Jermaine's speed?

Get ConnectED *For more practice, go to* www.connected.mcgraw-hill.com.

Multi-Part Lesson 1

PART E

Problem-Solving Practice

Convert Rates

DRIVES Use the table for Exercises 1–2. The table shows the average speeds for drives to the beach.

Average Speeds	
Name	**Speed**
Julie	62 miles per hour
Manalo	311,520 feet per hour
LeShawn	1,448 meters per minute

1. Find LeShawn's average speed in meters per hour.

2. List the names and corresponding speeds from least to greatest in miles per hour.

3. VEGETABLES Marilyn ate 320 pounds of vegetables last year. How many ounces did she eat per month? Round to the nearest tenth.

4. ANIMALS A chicken can run at a top speed of 9 miles per hour. Find the top speed of a chicken in miles per minute. Round to the nearest tenth.

5. COYOTES The top speed of a coyote is 43 miles per hour. Find the approximate speed in kilometers per minute.

6. WALK Heather was out for a leisurely walk at a rate of 3 miles per hour. What was her speed in yards per minute?

7. POOL A pool is being drained at a rate of 120 gallons per minute. What is this rate in quarts per hour?

8. MILK The Prichard family drinks 2 quarts of milk per day. How many gallons of milk do they drink in a week?

Course 2 • Measurement and Proportional Reasoning

Multi-Part Lesson 1

PART F

Homework Practice
Convert Units of Area and Volume

Complete. Round to the nearest hundredth if necessary.

1. $17,000,000 \text{ cm}^3 = $ _____ m^3 2. $2 \text{ yd}^2 = $ _____ in^2

3. $0.25 \text{ m}^3 = $ _____ mL 4. $24 \text{ cm}^3 = $ _____ mL

5. $100 \text{ ft}^3 = $ _____ yd^3 6. $3,600 \text{ in}^3 = $ _____ ft^3

7. $0.65 \text{ m}^2 = $ _____ cm^2 8. $13 \text{ ft}^2 = $ _____ yd^2

9. $710 \text{ in}^2 = $ _____ ft^2 10. $81 \text{ yd}^2 = $ _____ ft^2

11. **SHOEBOX** A shoebox has a length of 12 inches, a width of 7 inches, and a height of 5 inches. Find the volume of the shoebox in cubic feet.

12. **EXERCISE** An exercise floor has an area of 40 square meters. Find the area in square centimeters.

13. **LAKES** The largest lake in Florida is Lake Okeechobee with an area of 700 square miles. If one mile is equivalent to about 1.61 kilometers, what is the area of Lake Okeechobee in square kilometers?

PETS Use the table at the right.

14. What is the volume of Spike's dog house in cubic feet?

15. Compare the volumes of Rex's dog house and Spot's dog house in cubic feet.

Dog House	
Dog	**Volume**
Rex	25 ft^3
Spot	1.2 yd^3
Potato	32 ft^3
Spike	$41,472 \text{ in}^3$

16. Find the volume of Potato's dog house in cubic inches.

Get ConnectED *For more practice, go to* www.connected.mcgraw-hill.com.

Problem-Solving Practice

Convert Units of Area and Volume

1. **PLAY AREA** Josam wanted to enclose an area of his yard for his children to play in. The dimensions of the enclosed area are 20 feet by 30 feet. How many square yards is this? Round your answer to the nearest tenth if necessary.

2. **GIFT BOX** Find the number of cubic inches that this gift box will hold.

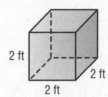

2 ft
2 ft
2 ft

3. **REFRIGERATOR** The Vargas family bought a new refrigerator with a volume of 23,328 cubic inches. Find the volume in cubic feet.

4. **CHECKERBOARD** A checkerboard is 8 squares by 8 squares. The squares on the checkerboard Mikal was using measured 2 centimeters on each side. What is the area of the checkerboard in square millimeters?

5. **LAWN MOWING** The lawn Jarron mowed was 40 yards by 65 yards. What is the area of the lawn in square feet?

6. **MEDICINE** Rahm is taking medicine that requires him to take 10 milliliters twice a day. How many cubic centimeters of medicine does he take per day?

Homework Practice

Problem-Solving Investigation: Make a Model

Mixed Problem Solving

For Exercises 1 and 2, use the *make a model* strategy to solve the problem.

1. **ARCHITECT** Mrs. Peron is designing a home for a client. The house is 45 feet by 76 feet. If she uses a scale of $\frac{1}{2}$ inch = 1 foot, what are the dimensions of the house on the blueprints?

2. **SWIMMING POOL** Mr. Forrester has a swimming pool that measures $3\frac{1}{3}$ yards by 8 yards. If the deck around the pool is $2\frac{2}{3}$ yards wide, what is the outside perimeter of the deck?

Use any strategy to solve Exercises 3 through 6.

3. **BATTERIES** A manufacturing plant can make 350 batteries in 15 minutes. How long will it take the manufacturing plant to make 3,500 batteries?

4. **SHOPPING** A grocery store has five cash registers. About 4 customers are checked out at each register every 20 minutes. How many customers are checked out at the store each hour?

5. **TESTS** Diego scored a 95 on his first test in science class. He then scored 100 on his next 5 tests. If he scored a 91 on his seventh test, what is his test average?

6. **NEWSPAPERS** Candace wants to increase the number of newspapers she delivers. She currently delivers 58 newspapers. In fourteen weeks, she wants to be delivering 100 newspapers. How many newspaper deliveries must she increase each week to obtain her goal?

Get ConnectED *For more practice, go to* www.connected.mcgraw-hill.com.

Problem-Solving Practice

Problem-Solving Investigation: Make a Model

Solve each problem using any strategy you have learned.

1. FOOTBALL Bill, Dakeem, and Hans are the quarterback, center, and punter on the football team, not necessarily in that order. The quarterback and Bill go on the bus with Dakeem after the game. Dakeem is not the punter. What position does Bill play?

2. SPORTS Tennille can walk one mile in 15 minutes. How long will it take her to walk 3 miles?

3. WEATHER The Loudonville Times prints the following chart showing the snowfall for each day last week. The reporter estimates that they got 10 inches of snow during the past week. Is this a reasonable estimate?

Day	Snowfall
Monday	1 inch
Tuesday	2 inches
Wednesday	0.5 inch
Thursday	1.5 inches
Friday	3.75 inches
Saturday	0 inches
Sunday	0 inches

4. GARDENING The table below shows how many tomatoes Nicholas picked each day during the week. How many does he need to pick on Sunday so that he has picked a total of 20 for the week?

Day	M	T	W	R	F	S	S
Number of Tomatoes	2	5	3	1	0	5	

5. PAINT If one gallon of paint covers 150 square feet, is one gallon enough for Fumiko to cover a kitchen wall that is 15 feet by 8 feet? Justify your answer.

6. SHOPPING Avery bought a DVD for $22.99 and received $2.01 in change. How much did Avery give the cashier?

7. MONEY The amount in Carly's checkbook is $750 after writing a check for $65 and making a deposit of $100 and a deposit of $75. How much did she start with in her checkbook?

8. VEHICLES Chang-Hee has 15 vehicles at his garage. Some are cars and some are motorcycles. If he counts 58 wheels, how many of each type of vehicle does he have?

Homework Practice

Changes in Dimensions

1. The surface area of a cube is 400 square millimeters. What is the surface area of a similar cube that is larger by a scale factor of 3?

2. **CANDLES** The volume of a candle is 8 cubic inches. What is the volume of a similar candle that is larger by a scale factor of 1.5?

3. **TRAVEL** The volume of a suitcase is 4.2 cubic feet. What is the volume of a suitcase that is smaller by a factor of 0.9? Round to the nearest tenth.

4. **DELI** A deli owner uses 215 square centimeters of plastic wrap to cover a wedge of cheese. How many square centimeters of plastic wrap would she need to cover a wedge of cheese with a similar shape that is smaller by a scale factor of $\frac{1}{2}$? Round to the nearest tenth.

5. **CRACKERS** A box of crackers has a volume of 48 cubic inches. What is the volume of a similar box that is smaller by a scale factor of $\frac{2}{3}$?

6. The surface area of a pyramid is 88 square feet.

   a. What is the surface area of a similar pyramid that is larger by a scale factor of 5?

   b. What is the surface area of a similar pyramid that is larger by a scale factor of 8?

   c. What is the surface area of a similar pyramid that is smaller by a scale factor of $\frac{1}{10}$? Round to the nearest tenth.

7. A cylinder was enlarged by a scale factor of 4. The new volume is 2,240 cubic units. What was the volume of the original cylinder?

 *For more practice, go to* www.connected.mcgraw-hill.com.

Problem-Solving Practice

Changes in Dimensions

PACKING Use the table for Exercises 1–3. The table shows the volumes of three types of packing boxes offered by a moving company.

Volume of Packing Boxes, in³	
Type A	5,000
Type B	7,500
Type C	10,000

1. Taso needs a box that is similar to Type A but that is larger by a scale factor of 2.5. What would be the volume of this box?

2. Kristina needs a box that is similar to Type C but is smaller by a factor of $\frac{1}{2}$. What would be the volume of this box?

3. The moving company used to offer Type D, which was similar in shape to Type B, but was larger by a scale factor of 3. What was the volume of Type D?

4. **DECORATION** Odell had a cone-shaped decoration on her dresser. It has a volume of 6,800 cubic millimeters. What is the volume of a similar cone that is $\frac{1}{5}$ this size?

5. **BIRD CAGE** Buan built a bird cage with a surface area of 540 square inches. Her sister Sirib built a bird cage with a similar shape, and it is larger than Buan's bird cage by a scale factor of 2.25. What is the surface area of Sirib's bird cage? Round to the nearest tenth.

6. **DETERGENT** For a limited time, a brand of detergent is being sold in a larger size for the same cost as the original size. The two boxes are similar in shape. The surface area of the original box is 1,200 cubic centimeters and the surface area of the larger box is 2,028 cubic centimeters. How much greater is the height of the larger box than the original box?

Multi-Part Lesson **1**

PART **A**

Homework Practice

Angle Relationships

Name each angle in four ways. Then classify the angle as *acute*, *obtuse*, *right*, or *straight*.

1.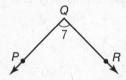

2.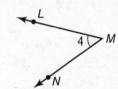

Identify each pair of angles as *complementary*, *supplementary*, or *neither*.

3.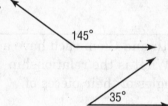

4.

5.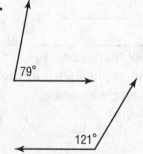

6.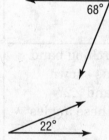

7. **ALGEBRA** If $\angle A$ and $\angle B$ are supplementary and the measure of $\angle A$ is 124°, what is the measure of $\angle B$?

8. **ALGEBRA** If $\angle X$ and $\angle Y$ are complementary and the measure of $\angle X$ is 85°, what is the value of $\angle Y$?

Get ConnectED *For more practice, go to* www.connected.mcgraw-hill.com.

Problem-Solving Practice

Angle Relationships

1. STREETS What type of angle is formed by Marquette Street and Garfield Avenue?

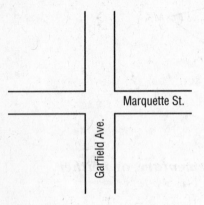

Marquette St.

Garfield Ave.

2. AIRPLANE An airplane flies due north and then turns northeast as shown. What type of angle is formed by the two routes?

3. GATE A gate has a diagonal support. Describe the relationship between ∠1 and ∠2.

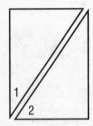

1
2

4. PIZZA Mary Beth and Clem each have a piece of pizza. What is the relationship between the angles of their pieces of pizza?

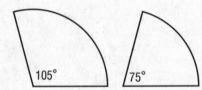

105°   75°

5. MARCHING BAND The marching band formed an X on the field as shown. What name refers to ∠1 and ∠2? How do the measures of these angles compare?

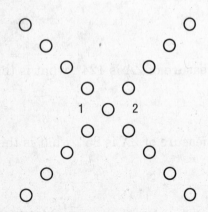

1   2

6. SOFTBALL The center fielder had to cover the area of the angle shown in the figure. Classify the angle as *acute*, *obtuse*, *right*, or *straight*.

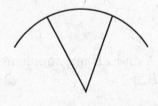

Multi-Part Lesson 1

PART C

Homework Practice

Triangles

Find the value of x.

1.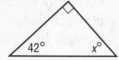

2.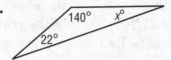

3.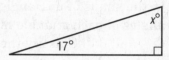

Find the missing measure in each triangle with the given angle measures.

4. 45°, 35.8°, x°

5. 100°, x°, 40.7°

6. x°, 90°, 16.5°

7. Find the third angle of a right triangle if one of the angles measures 24°.

8. What is the third angle of a right triangle if one of the angles measures 51.1°?

9. ALGEBRA Find $m\angle A$ in $\triangle ABC$ if $m\angle B = 38°$ and $m\angle C = 38°$.

10. ALGEBRA In $\triangle XYZ$, $m\angle Z = 113°$ and $m\angle X = 28°$. What is $m\angle Y$?

Classify the marked triangle in each object by its angles and by its sides.

11.

12.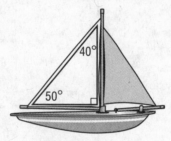

13.

ALGEBRA Find the value of x in each triangle.

14.

15.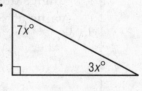

16.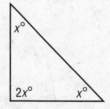

Get ConnectED *For more practice, go to* www.connected.mcgraw-hill.com.

Problem-Solving Practice

Triangles

1. TAILORING Each lapel on a suit jacket is in the shape of a triangle. The three angles of each triangle measure 47°, 68°, and 65°. Classify the triangle by its angles.

2. FLAGS A naval distress signal flag is in the shape of a triangle. The three sides of the triangle measure 5 feet, 9 feet, and 9 feet. Classify the triangle by its sides.

3. CARPENTRY The supports of a wood table are in the shape of a right triangle. Find the third angle of the triangle if the measure of one of the angles is 23°.

4. MAPS The three towns of Ripon, Sparta, and Walker form a triangle as shown below. Classify the triangle by its angles and by its sides. What is the value of x in the triangle?

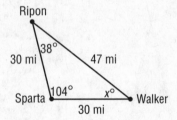

5. HIKING The figure shows the Oak Creek trail, which is shaped like a triangle. Classify the triangle by its angles and by its sides. What is the value of x in the figure?

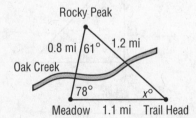

6. LADDER The figure shows a ladder leaning against a wall, forming a triangle. Classify the triangle by its angles and by its sides. What is the value of x in the figure?

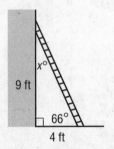

Multi-Part
Lesson **1**
PART **D**

Homework Practice

Quadrilaterals

Find the value of *x* in each quadrilateral.

1.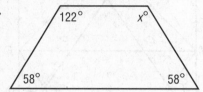
122° *x*°
58° 58°

2.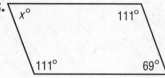
x° 111°
111° 69°

3.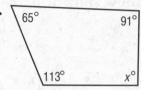
65° 91°
113° *x*°

4.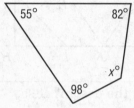
55° 82°
x°
98°

5.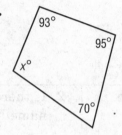
93°
95°
x°
70°

6.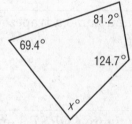
115° *x*°
110°
45°

7.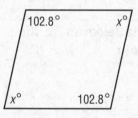
102.8° *x*°
x° 102.8°

8.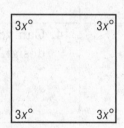
3*x*° 3*x*°
3*x*° 3*x*°

9.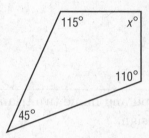
81.2°
69.4°
124.7°
x°

10. **FLAGS** Refer to the flag of Kuwait. Classify the quadrilaterals in the flag design.

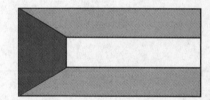

Classify each quadrilateral.

11.

12.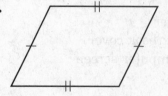

Get ConnectED *For more practice, go to* <u>www.connected.mcgraw-hill.com.</u>

Course 2 • Polygons and Transformations

NAME _____ DATE _____ PERIOD _____

Problem-Solving Practice
Quadrilaterals

STAINED GLASS Use the design for a square stained
glass window shown.

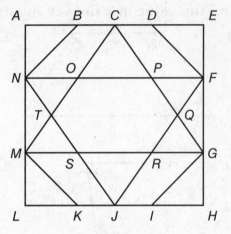

1. Find and name two triangles in the design.	**2.** Is there a square in the design? If so, name it.
3. Find and name a trapezoid in the design.	**4.** Can you find a parallelogram in the design? If so, name it.
5. Is the figure *CQRST* a quadrilateral? Explain.	**6.** If the perimeter of the window is 8 feet, what is the length of each side? How do you know?

COMMON OBJECTS Use the list of figures you see on a daily basis.

door
textbook cover
computer screen

stop sign
CD case

7. Which object on the list is *not* a quadrilateral?	**8.** Are there any objects on the list that are quadrilaterals? If so, what are they?

Homework Practice

Polygons and Angles

Determine whether each figure is a polygon. If it is, classify the polygon and state whether it is regular. If it is *not* a polygon, explain why.

1.

2.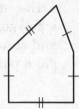

3.

4.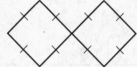

5.

6.

Find the sum of the measures of the angles in each polygon.

7. dodecagon
(12-sided)

8. 16-gon

9. 36-gon

10. nonagon

Find the value of each variable.

11.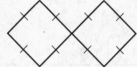

12.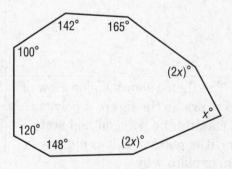

13.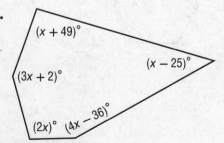

14.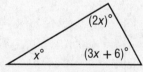

Get ConnectED *For more practice, go to* www.connected.mcgraw-hill.com.

Course 2 • Polygons and Transformations

189

Multi-Part 1
Lesson
PART E

Problem-Solving Practice

Polygons and Angles

1. ROYALTY The outline of a crown worn by a king is shown below. Is the figure a polygon? If it is, classify the polygon and state whether it is regular. If it is *not* a polygon, explain why.

2. ALCHEMY The symbol shown is one of the signs for *salt alkali* used in 17th-century chemistry. Is the symbol a polygon? If it is, classify the polygon and state whether it is regular. If it is *not* a polygon, explain why.

3. JEWELRY The symbol shown is often used to represent gems. Is the symbol a polygon? If it is, classify the polygon and state whether it is regular. If it is *not* a polygon, explain why.

4. SYMBOLS The 5-pointed star shown has sides of equal length. Is the symbol a polygon? If it is, classify the polygon and state whether it is regular. If it is *not* a polygon, explain why.

5. STAIRS The figure shows a side view of a set of stairs. Is the figure a polygon? If it is, classify the polygon and state whether it is regular. If it is *not* a polygon, explain why.

6. RUG The rug shown is a regular hexagon. What is the measure of each angle?

Course 2 · Polygons and Transformations

Homework Practice

Translations in the Coordinate Plane

1. Translate rectangle *ABCD* 3 units right and 4 units down. Graph rectangle *A'B'C'D'*.

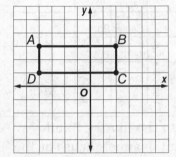

2. Triangle *PQR* is translated 3 units left and 3 units down. Then the translated figure is translated 6 units right. Graph the resulting triangle.

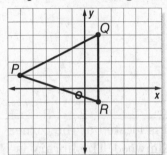

Triangle *EFG* has vertices *E*(1, 1), *F*(4, −3), and *G*(−2, 0). Find the vertices of *E'F'G'* after each translation. Then graph its translated image.

3. 3 units left, 2 units down

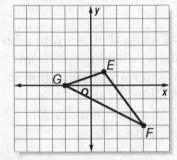

4. 4 units up

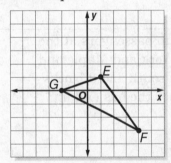

5. SEATS Jatin was given a new seating assignment in science class. The diagram shows his old seat and his new seat. Describe this translation in words and as an ordered pair.

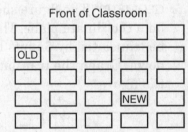

Front of Classroom

OLD

NEW

REASONING The coordinates of a point and its image after a translation are given. Describe the translation in words and as an ordered pair.

6. $A(1, -2) \rightarrow A'(3, 4)$

7. $H(3, 3) \rightarrow H'(-4, 0)$

8. $Z(-2, -4) \rightarrow Z'(1, -5)$

Get ConnectED *For more practice, go to* www.connected.mcgraw-hill.com.

Multi-Part Lesson **2**

PART **B**

Problem-Solving Practice

Translations in the Coordinate Plane

MAPS For Exercises 1–4, use the map at the right.

		Maple St.			N
		Blonde St.			
		Dodge St.			
		Pacific Ave.			
		Center Rd.			
		Harrison St.			

Kensington Ave.   Elmwood Ave.   Delaware Ave.   Main St.   New York Ave.   California Ave.

---

1. Stanley's school is located at the corner of Center and Elmwood. The library is located at the corner of Dodge and Delaware. Describe Stanley's walk from school to the library as an ordered pair of the number of blocks.

2. After he goes to the library, Stanley goes to his Aunt Jeanne's house at the corner of California and Harrison. Describe Stanley's walk from the library to his aunt's house as an ordered pair of the number of blocks.

3. If a bus picks up passengers at the corner of New York and Maple and drives 2 blocks south and 3 blocks west, where does the bus end up?

4. Organizers of a walkathon want to map out a route that will lead people from the corner of Center and Kensington to the corner of California and Maple. Write a coordinate pair that describes the most direct route.

5. GEOMETRY The figure shows an octagon on a coordinate plane. The figure is to be translated by 5 units left and 5 units down. Graph the translated image of the figure.

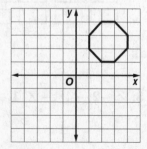

6. BANKS Clarissa is waiting in line at the bank. There are several people in line in front of her. Describe the path Clarissa must take to get to the front of the line if each time she moves up in line by one position is considered one unit.

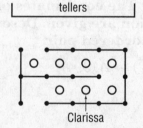

tellers

Clarissa

Homework Practice

Reflections in the Coordinate Plane

Draw the image of the figure after a reflection across the given line.

1.

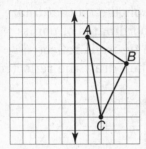

2.

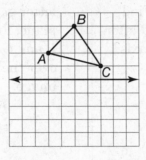

3.

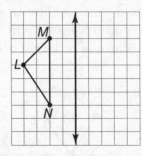

4.

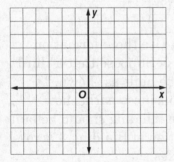

5.

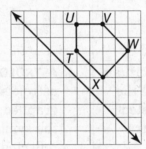

6.

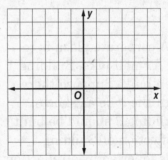

7. Graph △*ABC* with vertices *A*(2, 2), *B*(5, 4), and *C*(5, 1) and its reflection over the *x*-axis. Then find the coordinates of the reflected image.

8. Graph square *ABCD* with vertices *A*(−1, 2), *B*(2, −1), *C*(5, 2), and *D*(2, 5) and its reflection over the *y*-axis. Then find the coordinates of the reflected image.

The coordinates of a point and its image after a reflection are given. Describe the reflection as over the *x*-axis or *y*-axis.

9. $B(1, -2) \rightarrow B'(1, 2)$

10. $J(-3, 5) \rightarrow J'(-3, -5)$

11. $W(-7, -4) \rightarrow W'(7, -4)$

Get ConnectED *For more practice, go to* www.connected.mcgraw-hill.com.

Course 2 · Polygons and Transformations

Multi-Part Lesson **3** PART **B**

Problem-Solving Practice

Reflections in the Coordinate Plane

1. ALPHABET The figure shows the letter *V* plotted on a coordinate system. Find the coordinates of points *C* and *D* after the figure is reflected over the *y*-axis.

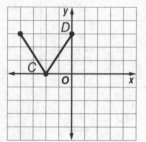

2. GREEK The figure shows the Greek letter *gamma* graphed on a coordinate plane. Find the coordinates of points *P* and *Q* after the figure is reflected over the *x*-axis. Then draw the reflected image.

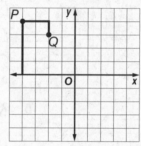

3. CRAFTS Daneen is making a pattern for star-shaped ornaments. Complete the pattern shown so that the completed star has a vertical line of symmetry.

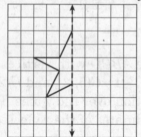

4. ASTROLOGY The figure shows the astrological symbol for Sagittarius graphed on a coordinate plane. Reflect the symbol across the *x*-axis. Graph the reflected image.

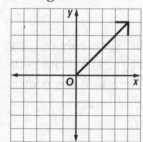

5. ARCHITECTURE A corporate plaza is to be built around a small lake. Building 1 has already been built. Suppose there are axes through the lake as shown. Show where Building 2 should be built if it will be a reflection of Building 1 across the *y*-axis followed by a reflection across the *x*-axis.

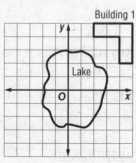

6. ARCHITECTURE Use the information from Exercise 5. Suppose that a third building is to be built as shown. To complete the business park, show where a fourth building should be built if it is a reflection of Building 3 across the *x*- and *y*-axes.

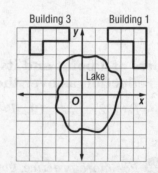

Multi-Part Lesson 4

PART B

Homework Practice

Rotations in the Coordinate Plane

1. Rotate △ABC 90° clockwise about the origin. Graph △A'B'C'.

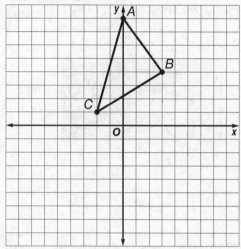

2. Rotate △ABC 180° clockwise about the origin. Graph △A'B'C'.

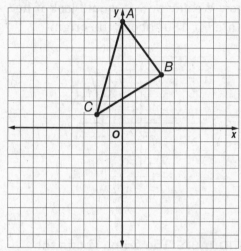

3. Rotate △XYZ 270° clockwise about the origin. Graph △X'Y'Z'.

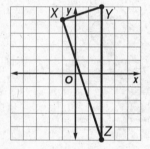

4. Rotate △XYZ 180° counterclockwise about the origin. Graph △X'Y'Z'.

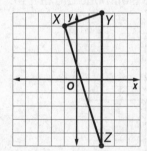

5. **SOFA** Quadrilateral ABCD represents the sofa in Lemont's family room. He would like to rotate it 90° clockwise about the origin. What will be the coordinates of the image?

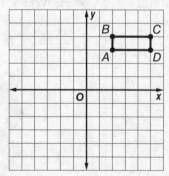

Get ConnectED *For more practice, go to* www.connected.mcgraw-hill.com.

Multi-Part Lesson **4**

PART **B**

Problem-Solving Practice

Rotations in the Coordinate Plane

1. POND The triangle represents the placement of a pond in Wesley's yard. Show Wesley how it would look if he rotated it 180° about the origin.

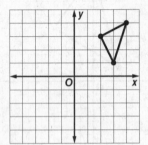

2. LOGOS Does the logo shown have rotational symmetry? If yes, what is the angle of rotation?

3. PATIO TABLE The graph shows the location of the patio table on Liam's deck. Show how it would look rotated 90° counterclockwise about the origin.

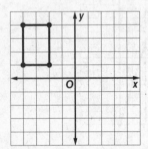

4. BUTTERFLIES Does the butterfly shown have rotational symmetry? If yes, what is the angle of rotation?

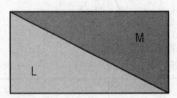

5. Does the logo shown have rotational symmetry? If yes, what is the angle of rotation?

6. FLOWERS Does the flower shown have rotational symmetry? If yes, what is the angle of rotation?

Multi-Part Lesson 5

PART A

Homework Practice

Dilations

Draw the image of the figure after the dilation with the given center and scale factor.

1. center: C, scale factor: 2

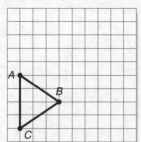

2. center: N, scale factor: $\frac{1}{2}$

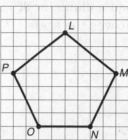

Find the coordinates of the vertices of polygon $F'G'H'J'$ after polygon $FGHJ$ is dilated using the given scale factor. Then graph polygon $FGHJ$ and polygon $F'G'H'J'$.

3. $F(-2, 2)$, $G(2, 3)$, $H(3, -2)$, $J(-1, -3)$; scale factor $\frac{3}{4}$

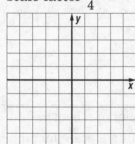

4. $F(-2, 2)$, $G(2, 4)$, $H(3, -3)$, $J(-4, -4)$; scale factor 2

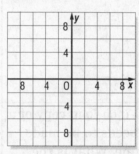

In the exercises below, figure $R'S'T'$ is a dilation of figure RST and figure $A'B'C'D'$ is a dilation of figure ABCD. Find the scale factor of each dilation and classify it as an enlargement or as a reduction.

5.

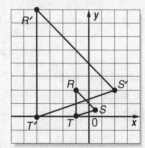

6.

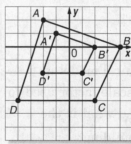

7. GLASS BLOWING The diameter of a vase is 4 centimeters. If the diameter increases by a factor of $\frac{7}{3}$, what will be the new diameter?

Get ConnectED *For more practice, go to* www.connected.mcgraw-hill.com.

Problem-Solving Practice

Dilations

1. EYES Dave's optometrist dilated his eyes. Before dilation, his pupils had a diameter of 4.1 millimeters. After dilation, his pupils had a diameter of 8.2 millimeters. What was the scale factor of the dilation?

2. BIOLOGY A microscope increases the size of objects by a scale factor of 8. How large will a 0.006 millimeter paramecium appear?

3. PHOTOGRAPHY A photograph was enlarged to a width of 15 inches. If the scale factor was $\frac{3}{2}$, what was the width of the original photograph?

4. MOVIES Film with a width of 35 millimeters is projected onto a screen where the width is 5 meters. What is the scale factor of this enlargement?

5. PHOTOCOPYING A 10-inch-long copy of a 2.5-inch-long figure needs to be made with a copying machine. What is the scale factor?

6. MODELS A scale model of a boat is going to be made using a scale factor of $\frac{1}{50}$. If the original length of the boat is 20 meters, what is the length of the model?

7. MODELS An architectural model is 30 inches tall. If the scale factor used to build the model is $\frac{1}{120}$, what is the height of the actual building?

8. ADVERTISING An advertiser needs a 4-inch picture of a 14-foot automobile. What is the scale factor of the reduction?

Homework Practice

Problem-Solving Investigation: Work Backward

Use the *work backward* strategy to solve Exercises 1–3.

1. RECTANGLE Rectangle $D'E'F'G'$ is a reflection over the *x*-axis. Find the original coordinates of the rectangle.

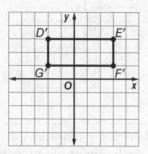

2. REFLECTION Suppose the rectangle in Exercise 1 was reflected over the *y*-axis. Would the ordered pairs for D' and E' and the ordered pairs for F' and G' switch locations?

3. DILATION The coordinates of the image of triangle *JKL* are $J'(4, 6)$, $K'(4, 2)$, and $L'(2, 2)$ after a dilation with a scale factor of 2. Find the coordinates of triangle *JKL* before the dilation and graph the triangle.

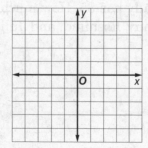

Use any strategy to solve Exercises 4–7.

4. A triangle with vertices $A'(2, 4)$, $B'(3, 1)$, $C'(-1, 2)$ was translated 3 units right and 2 units up. What are the coordinates of the original triangle?

5. Anton has pennies, nickels, and quarters in his pocket. He has $0.34. How many of each coin does he have?

6. Describe the translation that will move $X(-3, 4)$ to $(2, 5)$. Then use this translation to find the image of $Y(1, 6)$.

7. Mr. Simpson's breakfast bill was $11.08. He gave the waitress a ten-dollar bill and a five-dollar bill. From the change he received, he left 3 one-dollar bills and 3 quarters for a tip. How much money remained?

Get ConnectED *For more practice, go to* www.connected.mcgraw-hill.com.

Problem-Solving Practice

Problem-Solving Investigation: Work Backward

1. REFLECTION Clyde reflected point L over the y-axis to get $L'(4, -5)$. Find the original coordinates of point L.	**2. ROTATION** The triangle was rotated 180° clockwise. Find the original coordinates of the triangle. 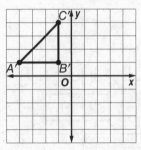
3. SWIMMING POOL Peter went swimming in a rectangular pool as shown. How many meters long is the pool?  4 meters Perimeter = 26 meters	**4. RIGHT TRIANGLE** Samantha drew a right triangle with one angle measuring 25°. What is the measure of the other angle?
5. TRANSLATION The ordered pair $(15, -8)$ was the result of a translation of 5 units up and 2 units left. Find the original coordinates of the ordered pair.	**6. MONEY** Julianne has $4 more than Rosalind but $3.50 less than Jenny. Jenny has twice as much money as Sally. Sally has $5. How much money does Jenny, Rosalind, and Julianne have?